THE Northlands
Winter Greenhouse
MANUAL

Publication of this book was made possible in large part by funding from the University of Minnesota West Central Partnership.

Additional Praise

It can be done! We have been eagerly awaiting the publication of this book to further our goal of creating a sustainable, year-round local food supply in North Dakota. **This no-nonsense manual will be our key to success.**
> — Sue B. Balcom, Local Foods Marketing Specialist
> for the North Dakota Department of Agriculture

When I first visited Garden Goddess on a bone-hard evening in January, I knew I was entering something special: a reminder of how ingenuity, grit, and a love of food can produce a verdant "door into summer" in the midst of frozen fields. This manual contains the seeds for making this passive solar greenhouse system not so unique after all, and that's **the key to creating a widely dispersed, sustainable food system**.
> — Brian DeVore, Editor,
> *The Land Stewardship Letter*

THE Northlands
Winter Greenhouse
MANUAL

A Unique,
Low-Tech
Solution to
Vegetable
Production
in Cold
Climates

Carol Ford &
Chuck Waibel

Garden
Goddess
publications
Milan, Minnesota

The Northlands **Winter Greenhouse** MANUAL
A Unique, Low-Cost Solution to Vegetable Production in Cold Climates

Published by Garden Goddess Publications, a division of Garden Goddess Enterprises.

Cover photo: Kristi Link Fernholz
Cover and book design: Ann Delgehausen, Trio Bookworks

Photography: Kristi Link Fernholz

The pictures on pages 15, 16 (top two), 18, and 21 (left) were taken by Chuck Waibel.

978-0-615-29724-8

Manufactured in Canada.
13   12   11   2   3   4   5   6   7   8   9   10

You are welcome to contact Chuck and Carol
to have them speak at your event or consult for your project.

newworld@fedteldirect.net *or* **320.734.4669**
www.gardengoddessenterprises.com

To MY DEAR FAMILY, even though, sadly, some of you still hate vegetables, thanks for always being there, believing in me, and offering unconditional love and support. I will continue in my efforts to convert you.

—Carol

To ROBERT HEINLEIN, KIM STANLEY ROBINSON, LARRY NIVEN, and the rest of the great writers who showed us that our future can be glorious, even if the present looks rocky—provided that we use the brains and insights God gave us.

—Chuck

NOTE TO OUR READERS

Paralleling the general division of our labor in raising the greenhouse structure and produce is the division of our writing in this book. We wrote the Beginnings section together and the other sections according to our expertise, hence the two distinct first-person voices.
Welcome to the Garden Goddess greenhouse!
—Carol & Chuck

WEST CENTRAL MINNESOTA

REGI●NAL
SUSTAINABLE DEVELOPMENT
PARTNERSHIP

Communities and their University building Minnesota's future

The University of Minnesota West Central Partnership,
legislatively funded in 1999, serves twelve counties of the
Upper Minnesota River Valley watershed. This citizen-led
initiative's mission is launching education and research that
help people understand and achieve sustainability in their region.
Since its inception, West Central Partnership has funded/partnered
on more than fifty applied research and education projects
in four priority areas: community-based clean/renewable energy,
local foods/sustainable agriculture, sustainable built/natural
resources, and building capacity in citizens and communities.
www.regionalpartnerships.umn.edu.

Contents

GROWING THE PRODUCE

LEARNING FROM EXPERIENCE

Ken Meter, president of Crossroads Resource Center in Minneapolis, has compiled more than forty-five regional farm and food economy studies in twenty states and one Canadian province. He frequently speaks on the topic "Local Foods as Economic Recovery" and related themes.

Foreword
Ken Meter

My first visit to Garden Goddess Greenhouse was on a crisp January day. A biting wind penetrated to my bones, and my eyes teared up. I had to lean into the gusts to keep my balance. The wind chill measured 13 degrees below zero. I wondered how the delicate plants might be faring inside Carol Ford and Chuck Waibel's backyard structure.

By the time Chuck led me into the greenhouse, the sun was sliding to the horizon. Moist, warm air beckoned me in. Fresh odors of soil and sprouting plants filled my nostrils. Glistening leaves assured me the plants were thriving. Although the wind still howled outside, the structure stood serene, with nary a rattle of complaint. The thermometer on the inside wall read 65 degrees. Inside this snug oasis, I immediately unzipped my down jacket.

In fact, Chuck boasted, "We had to open the greenhouse door for an hour or so today, so it wouldn't be too hot for you. It was up to 85 degrees in here at mid-afternoon."

Hanging from the rafters were dozens of household gutters, each filled with a rich black soil. Multicolored salad greens stood upright inside each gutter garden plot. Larger crops like broccoli grew in the dirt floor of the greenhouse, glowing a rich green. There was a profound sense of order inside this space.

Most inspiring, the warm temperatures inside the greenhouse were due almost solely to the effect of the sun. Carol and Chuck had wisely constructed their growing space by digging four feet underground, well below the frost line, then placing gravel in this pit. Rich soil was piled on top of the rock. When the sun shone, the air inside was warmed, then diverted underground through a pipe. The subterranean gravel stored this heat. At night, or when the sun was covered by thick cloud layers, this warm air would rise into the greenhouse, keeping the space temperate. If it ever fell to 45 degrees inside, a propane heater kicked in to warm the space, keeping the plants from freezing. Chuck said the fuel for this heater costs about $50 per year.

Standing in this warm haven, viewing the snow banks outside through polycarbonate, the future seemed alive with possibilities. The taste of fresh, tangy mustard greens convinced me that building more greenhouses modeled after Carol and Chuck's is Minnesota's most essential strategy for extending our growing season.

Our state wrestles with profound ironies in agriculture. Although the seventh largest farm state in the United States, we import about 90 percent of our food. This means we ship $11 billion out of our state each year, purchasing food from distant sources. This would be nearly enough money to pay state farmers for all of the $13 billion of commodities they produce each year—yet these farm products are primarily used for industrial processing, not sold directly to humans.

Although our state's farmers supply global commodity markets, they have sparse connection to Minnesota kitchens and dinner tables. For example, direct sales of food by farmers to consumers totaled $35 million in 2007, only 0.3 percent of total farm product sales. These direct sales are rising rapidly, increasing 52 percent from 2002 to 2007, which is a sure sign of the hunger Minnesota consumers feel to know more about the sources of their food, and to make contact with farm families.

The lack of connection is so astounding that I once caught an Iowa farmer short by suggesting there was a solid market in his own rural region for selling food. His face twisted a bit, and he asked, incredulously, "You mean, if I raised food for people, someone would buy it?" It was a remarkable statement for a farmer to make in the midst of a nation that prides itself on "feeding the world."

Attempting to feed the world has come with certain costs. Minnesota farmers spent $7 billion more producing crops and livestock over the 14 years 1994–2007 than they earned by selling these products. Thirty-eight percent of Minnesota farms reported a net loss in 2007. Despite doubling productivity, farmers earned $1 billion less in 2007 than they had in 1969.

Moreover, the state's food system is deeply dependent on a most vulnerable resource: oil. Our willingness to ship food long distances, our ability to send large tractors and combines into the fields, and the logistics of importing over half of the fertilizer used on our massive grain fields—all of this boils down to assuming that oil will be (1) available and (2) cheap. As oil supplies peak, both assumptions are breaking down.

This makes Carol and Chuck's work all the more urgent. While shipping fresh greens to the twenty families who buy shares for deliveries from October to April makes but a small dent in the economic losses described above, the act of providing neighbors with quality organic produce is an essential act for rebuilding community, a sense of hope, and healthy eating in Minnesota.

If you have not yet met Carol and Chuck, you are likely to enjoy the way their humor erupts on the pages you are about to read. If you have met them, you will recall their twinkling eyes and firm faith in their Minnesota mission as you read this book. Anyone who cares about how Minnesotans will eat in a post-oil economy should read this book right away—and begin checking for sunny spots in their community. Policy wonks would also do well to study this practical manual themselves. They would pick up a few solid hints about gardening—as well as how to envision new social futures.

Chuck and Carol move in stark contrast to one of my hosts a few years ago, who sponsored a community forum where I could talk about local foods one winter, yet apologized rapidly when the subject of what to serve the guests came up. "We can't serve local foods in the winter," I was told in a firm voice, "after all, we are in Minnesota."

I held my tongue, but I was puzzled by this notion that residents of the seventh largest farm state in the United States had lost track of how to plan for winter months. Surely, this state grows wild rice, grains, potatoes, onions, beets, carrots, squash, and apples that could be stored for use in colder weather. Certainly, local meats can be smoked, or frozen, for later use. Clearly we have winter supplies of cheese, eggs, and dairy products. Many fruits and vegetables can be frozen or canned. Yet all of this requires planning and time, which have become scarce.

Now, thanks to Carol and Chuck, we can add the promise of savoring the flavors of fresh greens and broccoli, even as the blustery January winds blow across the Minnesota prairies.

Beginnings

When we became hooked on the fresh local vegetables provided through Easy Bean Farm, a CSA (community supported agriculture) farm, we learned to dread that first late autumn trip to the grocery for produce. Grim and depressed, we would pick up a bag of salad greens or a head of broccoli and wonder, Where was this grown? What did they spray on it? What was in the water that irrigated the field it grew in? How old is this vegetable, and who handled it on its way to me? And how much gasoline was burned getting this vegetable to my town?

So why don't we grow winter vegetables here in Minnesota? Because we don't think we can. Well, after four years of raising fresh vegetables during western Minnesota's fierce prairie winters, we are in love with what a little passive solar technology can do.

CSA

Community supported agriculture. A means of food distribution where consumers, or shareholders, agree to pay a fixed sum to a farmer at the beginning of the growing season and receive regular deliveries of harvested produce.

Imagine that it is mid-January and a blizzard blasts outside the safe comfort of your northern home. You are enjoying a bowl of fresh-picked salad greens with texture, flavor, and vibrant color. These salad greens provide you with maximum vitamin and mineral goodness. Better still, they are so delicious, you don't even bother with dressing. You snack on greens like you used to crunch potato chips. This is revolutionary.

You enjoy complete confidence in this locally grown food. There's no question as to where or how it was grown, because you did it! And if it's a warm meal you want on a cold winter night, you can always go for a stir fry of pac choi (bok choy) and broccoli fresh from your winter greenhouse. How did you ever get along without this winter survival kit?

Can't Be Done, Huh?

When we first started to explore the possibilities of winter vegetable production, we sought out the advice of experts in our region. We talked to organic vegetable farmers who have tried season extension with low-tech hoop house structures, following in the footsteps of author/farmer Eliot Coleman and his four-season gardening strategy. Coleman has some stellar ideas, including the ambition to always choose the simplest solution for winter production. The less reliant a system is on fossil fuels, the better suited it is to address our production needs now and in our challenging future.

high tunnel, hoop house

These terms are used interchangeably in most literature about protective structures used for vegetable production season extension. Both refer to a framework of hollow metal tubing covered by one or two layers of plastic sheeting. In most cases, these terms refer to structures that do not use heaters to raise the air temperature.

But Coleman's East Coast model doesn't quite work in a wind-beaten northern prairie climate. A hoop structure covered with one or two layers of plastic cannot keep soil and plants from freezing during a stretch of midwinter weather, with minus 40–degree F/C windchill, without the addition of costly supplemental heat. Imagine the effects of an Alberta clipper's gusty fifty-mile-an-hour winds on a structure of metal tubes and plastic sheeting, made brittle by winter's breath.

Grow greens in Minnesota in the winter? "Can't be done," the experts warned us. "Heating costs are prohibitive, and soil freezes up. Season extension is the best you can hope for."

Well, here's the deal. "You can't do it" hits our sensibilities like a challenge. We wanted to examine the answer to "Why not?" and see if we could come up with a solution. Why nots are just design factors, not project killers. So we decided to take a look at the possibilities along with the roadblocks.

The Long and Short of It

When we decided to write a book about the Garden Goddess greenhouse, we envisioned an all-encompassing collection of information and stories about our path from conception to execution. That book is still in the works, but something happened along the way that changed our priorities. More and more people were calling and writing to us, asking for the specifics on very practical aspects of our business: the design, features, and rationale of the structure as well as specific information regarding production during the winter months.

Garden Goddess Produce is a year-round business. In the winter, we offer a CSA with deliveries from mid-October to mid-April. In spring, we finish up our deliveries season, clean out the greenhouse, and tend seedlings that will be transplanted into the summer garden plots. Summer is for growing the outdoor crops for fall harvest that will wait in cold storage to be doled out during winter months. Fall gets crazy because we are planting in the greenhouse, harvesting outdoors, and packing and delivering shares each week.

If you picked up this manual because you are intrigued by the possibilities of a winter greenhouse, you may or may not

greenhouse

In this book the term "greenhouse" refers to the passive solar winter greenhouse; when needed for clarity, other kinds of greenhouses are referred to as "conventional."

have the same sort of business in mind that we have developed. You may have a good market at local gourmet restaurants for fresh winter herbs or a food co-op in your town that will happily sell all the greens you can grow. So with the multiple possibilities for a winter greenhouse in mind, we'll jump in at autumn time, when anyone with a winter greenhouse is going to be plenty busy getting all those baby plants going that will feed happy customers during the most challenging season of the Upper Midwest.

RAISING THE STRUCTURE

Designing and Building a Garden Goddess Passive Solar Winter Greenhouse

It's exciting when you experience one of those "head slapper" moments, when you realize that the answer to a problem was staring you in the face all along. Just by following a trail of thought, you end up becoming a pioneer, doing what no one has done before.

The Garden Goddess greenhouse is like that. In cold climates it's "obvious" that we can't feed ourselves in the winter. But that apparent truism is wrong. We can be pioneers, doing what couldn't be done before. To our knowledge *this is the only greenhouse of its kind, anywhere*—at least until *you* build another.

Even so, the design and construction of the Garden Goddess passive solar winter greenhouse is very straightforward. Some of the math seems a bit thick at first, but not once you get used to it. We'll talk about factors you need to know about for your own designing, then explain what we did. If you're building roughly along the line from Portland, Oregon, through Toronto to Portland, Maine, you can just copy what we did and come out fine, because your sun angle through the year is about the same as it is here in Milan, Minnesota. In warmer areas, such as along the coasts, you may need less insulation and more ventilation. In mountainous or windier regions, you'll need heavier insulation.

We can usually do more than we think we can. You can do this.

The Four Planning Considerations: Sun, Soil, Power, and Water

Sun

We've all heard in high school earth science class that summer days are longer than winter days, that the sun rises farther to the north and goes higher in the sky in summer than in winter. Most of us don't pay much attention to this, but if you work with solar energy you must pay close attention.

I remember when I first noticed this. When I was a boy, our family business of delivering newspapers had me up before dawn year-round. I couldn't help but notice the way sunlight changed with the seasons.

The changes in sun angle are important for two main reasons. First, when capturing sunlight for heat, the angle at which the light hits your glazing influences how much light you capture and how much is just reflected. You want this solar incidence angle to be as close as possible to straight out from your glazing at the time of year when you most need it: the depth of winter, around December 21. That's not usually the coldest time of year, just when the days are shortest. You'll need this sun height angle when figuring the slant of the south wall of your greenhouse. This is the only transparent wall, where you gather sunshine.

glazing

general term for glass, polycarbonate, polypropylene, or other clear or translucent wall covering

solar incidence angle

the angle at which sunlight strikes your south wall

The ideal angle is perpendicular to the sun's highest elevation on December 21. A solar path diagram will give you an exact figure, but as a general rule, take your latitude (how many degrees north of the equator you are), and subtract 23.5 degrees for how far down the earth's tilt depresses the sun in winter. Then subtract that number from 90 degrees.

90° north of the equator – Earth's tilt

= the ideal angle for your greenhouse south wall

A good atlas can show you this. For the Garden Goddess greenhouse, at 45 degrees North, that ideal figure is 45° + 23.5° = 68.5°; in Tulsa, that's 36° + 23.5° = 59.5°; in Saskatoon, 52° + 23.5° = 75.5°.

Second, how far north or south the sun rises and sets will influence how much of the day's sun hits your greenhouse. You'll want the maximum exposure in winter and the minimum in summer. Where the sun rises and sets will tell you much about how to site the greenhouse. Your view of the sky from southeast to southwest should be as unobstructed as possible. If there are obstructions, like buildings, trees, or towers, make sure that their shadows won't fall on your greenhouse.

The Garden Goddess greenhouse sits almost exactly at 45 degrees North. Minneapolis and St. Paul, the big cities in our region, are also near this line. On the equinoxes, about March 21 and September 21, the sun here rises and sets pretty much due east and west. It climbs to 45 degrees above the southern horizon at noon.

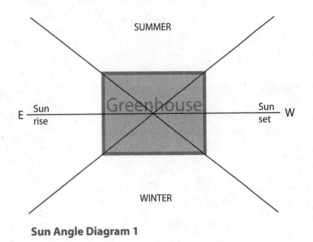

Sun Angle Diagram 1

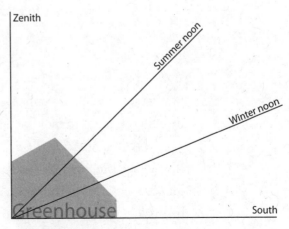

Sun Angle Diagram 2

On the summer solstice, about June 21, the sun climbs to about 68 degrees above the southern horizon at noon, and rises and sets 45 degrees north of due east and west. On winter solstice, about December 21, the sun climbs to only about 21 degrees above the southern horizon at noon, and rises and sets 45 degrees south of due east and west.

To get these numbers, which are the solar incidence angles for your location, you need to do a fairly complex calculation, or use a sun path diagram. These can be found in libraries, at solar energy equipment dealers, or online (see the Resources section at the back of the book).

sun path diagram

diagram that shows sunrise, sunset, and maximum sun height angles for a given latitude

Of course, these figures give the ideal solar incidence angle. Architectural concerns may force you to compromise. As a general rule, any angle within 25 degrees of ideal will still get you 90 percent of the incoming solar energy. For instance, we went with a 45-degree slant for the south wall, since this was close enough and much easier for our carpenters to work with. Carpenters prefer to work with right angles and 45-degree angles because that's how tools are designed.

Soil

Soil isn't just where your crops will grow. It's also the support for your entire structure. Use the same basic considerations you would for building a house. Here are some useful questions to ask yourself and your contractors:

- *Is the area stable?*
- *Is there good drainage?* Flooded footings can be as bad in a passive solar winter greenhouse as a flooded basement is in a dwelling, leading to mold problems.
- *How deep can, or should, I put the foundation?* For instance, if you have a looser soil, you'll need a broader, deeper foundation for stability. If you have only a few feet to bedrock, a shallower foundation will do. Local building codes and zoning regulations will often tell you how deep

your foundation should be for legal purposes. Be sure to check them early in your planning.

- *How deep is the average frost line, the depth to which the soil freezes in winter?* The frost line depth will tell you how deeply you must insulate. Below the frost line the soil temperature is generally around 50 degrees year-round. Not going below the frost line is like running an oven and leaving the door open—your passive solar winter greenhouse won't hold heat well.

- *Will I plant directly in the ground?* If so, find out if the soil is fertile and uncontaminated, or see if you can make it so.

These considerations will be different if you are building on an existing foundation or slab, or if you're putting the greenhouse on a roof. Wherever it goes, the underlayment must be able to handle the weight and to block out frost. For example, when we were asked to plan a passive solar winter greenhouse for a school cafeteria roof, our first consideration was whether the roof was strong enough to bear the weight of the greenhouse, the soil, the plants, and water. Fortunately, that school wing had been originally designed to have a second floor added when the district needed and could afford it, so we were in the clear.

If you're building on a parking lot or the slab of an old building, make sure that it's strong enough to hold what you're building, and that the slab extends several feet outward from your walls to insulate against frost.

Frost is your number one enemy: Lack of insulation on the ground lets it through. When in doubt, insulate.

Power

Although most of the energy used by your greenhouse will be free solar energy, you will need some supplements. You will need electricity for fans and work lamps, and a backup heat source for stretches that are especially cloudy and cold. In general, regular power-company electricity and heat from gas or propane are fine.

If you are so blessed by location, a wind generator is an ideal source for electrical power.

Propane or corn-burning furnaces are good heat sources, provided that they can be thermostatically controlled.

When wiring your greenhouse, the electrician will want to know the amperes for each circuit—that is, how much power you need in each area. He will use this to determine how thick the wires will be, the rating of the circuit breakers, and so forth. When you plan your greenhouse, think about all the things you will need electricity for—fans, vent blowers, work lights, germination heating pads, space heaters, refrigerators—and where they will go. Every piece of equipment comes with a draw amp figure, the amount of power the unit will use. Think of each electrical device as a faucet that can let a certain amount of water through—that's the draw amp figure. Your electrical circuit will need to be a big enough "pipe" for all the "faucets" you attach to it.

draw amp figure

the amount of power an electrical device uses

Figure out these numbers for everything you'll need (estimates are fine), add them up, and then add a healthy safety margin. A 20- to 30-amp circuit is typical, so you'll likely want several circuits. This is where a qualified electrical contractor is indispensable. The Garden Goddess greenhouse has four circuits: one for the packing shed, one for the wall shared by the greenhouse and the garage, and one for each end.

Water

The Garden Goddess greenhouse is blessed with good-quality, low-chemical, small town water. In some cities you will need to filter water or let it gas out in a holding tank to remove noxious chemicals that can damage plants. In any case, you'll need water for the plants and for washing harvested vegetables, and a good drain tied into the sewer lines or a septic system. The water leaving the greenhouse will be low in contaminants.

Well water is a good alternative, but due consideration must be given to purity and local regulations. Some of the recent *e. coli* scares were caused by contaminated well water. High nitrate water will also damage your plants, and contaminants such as heavy metals or chemicals are a serious problem.

Collected rainwater is good, but be sure that it isn't contaminated by the roof it comes off of or by any air pollution in your area. Your holding tank may grow algae, but that's not likely a problem if you keep it covered and skim out the growth. We keep fish in one of our outside tanks to eat weeds and mosquitoes.

Building a Passive Solar Winter Greenhouse from the Ground Up

When we designed the Garden Goddess greenhouse, we asked, "What if we were building this on Mars?" That may sound silly, but with the obvious exception of air pressure, building in western Minnesota isn't really much different from building on the Red Planet: both are cold and windy. January temperatures of minus 25 degrees F (−4 degrees C) with twenty-five-mile per hour winds are common. Conventional greenhouses need prodigious amounts of heat, and hoop houses either shatter or turn into enormous kites.

We attached the passive solar winter greenhouse to our garage to shield the northern side from the prevailing winds. This was important, saving a great deal of heat and giving us a way to load boxes of vegetables into the car without going outside.

Although we followed extreme weather design guidelines, we compromised where necessary according to two principles:

1. Make it easy to replicate.
2. Use off-the-shelf materials wherever possible.

Accordingly, almost all materials needed for construction will be in stock at your local lumber yard, hardware store, or home improvement center. The only major exception is the polycarbonate glazing and its mounting hardware, which we got from FarmTek (www.farmtek.com). Our glazing was expensive, but yours could be much cheaper if you can find a local source for a reasonable substitute material.

At the end of this section you'll find the working drawings (pages 25–29), the closest thing to blueprints we had for our project. We knew what we wanted to do and more or less how to do it. We showed these drawings to our contractors and let them advise us on the details. This led to intense conversations and much learning all around.

We worked with four main kinds of contractors:

- Masonry, for the foundation
- Plumbing, for the water and drain
- Electrical, for the wiring
- Carpentry, for the main structure

If you exactly copy what we built, it should work all right anywhere within a hundred miles north or south of 45 degrees North. But feel free to improvise! If you have sources for equivalent materials, use them. If your local weather or laws make our way impractical, change it. If you want to make the south wall angle closer to ideal, go for it. Most especially, if you have the benefit of experienced contractors to talk things over with, listen to them. Just keep in mind the basic design elements.

Thermal Mass and Foundation

The Garden Goddess greenhouse is heated mostly from the ground up. This requires a thermal mass, a material that gathers and holds heat, releasing it when it's needed. The primary thermal mass for the greenhouse is a thick layer of river wash rock—very large gravel the size of golf balls. This is better than finer gravel or sand because it allows the warm air to percolate through the material better.

thermal mass

a material that gathers and holds heat

Black stovepipe near the roof peak collects heat that an in-pipe blower fan forces under the ground through drain tile. The tile runs through the large gravel layer, heating it.

The foundation is of standard 8-inch cinder block, just like most house foundations. It's 4 feet deep (past the local frost line), with 3-inch blue foam insulation boards on the inside.

In the bottom of the excavated foundation pit we put back about a foot and a half of subsoil. Then we added about 8 inches of gravel and laid several interconnected loops of 4-inch perforated black drain tile as used in field drainage systems. On top of this we put another 8 inches of gravel. This created the primary thermal mass.

The picture shows the foundation after the first half of the gravel and the perforated pipes had been laid. In the front and rear left are two 6-inch vertical PVC pipes, which tie directly into our heat collectors via a row of black perforated T-joints along the east and west walls. From these Ts eight rows of perforated drain tile cross the floor. At the right rear you can see one of two "loose" vertical PVC pipes that rest in the gravel but aren't attached to the black pipes. Air goes down the attached pipes, into the drain tile. This filters through the rocks, heating them. Cooled air comes out the loose pipes. (We show how the heat is gathered and sent downward in the Ventilation and Supplemental Heat section, pages 20–21.)

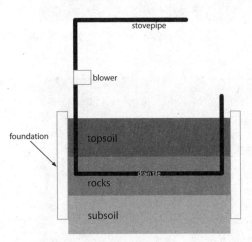

Heat Storage System

Foundation: The masons have nearly finished laying the blocks, and will then glue up the blue foam.

Perforated Pipes

Loose Exhaust Vents: The two loose exhaust vents and the west heat gathering vent stuck up as we began putting in the topsoil.

Water Barrels: Four water barrels do double duty as the secondary thermal mass and a convenient shelf.

On top of the gravel we put a layer of medium-weight white garden row cover as a sifting barrier. On top of this we put a foot of topsoil. The raised growing beds are a combination of black dirt, peat moss, mineral fertilizers, and compost. We grow broccoli, pak choi, Chinese cabbage, and other larger, slower-growing crops directly in this soil.

Along the back (north) wall we placed a row of four large plastic barrels filled with water. These comprise the secondary thermal mass. They act as a kind of shock absorber, slowing the rapid warming and cooling of the greenhouse at sunrise and sunset. If you put in more barrels, they will keep the greenhouse from getting as warm during the day or as cool at night, but there comes a break-even point where more barrels are counterproductive. In one smaller test unit we found that too many barrels kept the whole structure warm, but not warm enough, twenty-four hours a day. For this size greenhouse, four barrels is right. They also make a convenient shelf for germinating new plants.

Since you're unlikely to be building exactly the same size structure as we did, here are some of the volumes and ratios we used. You can adjust them as needed.

Interior Volume. Volume is height times width times length. For the Garden Goddess greenhouse, because of the 45-degree angles of the roof, this works out in two volume boxes, one for the bottom 4 feet from the front knee wall and one for the peak.

> Area: 16 x 22 = 352 square feet

> Volume (approximate): (352 x 6) + (352 x 4) = 352 x 10 = 3,520 cubic feet

Gravel Volume. The gravel is 1.5 feet deep across the entire area of the floor, hence:

> Volume of gravel: 352 x 1.5 = 528 cubic feet

Gravel is sold by the cubic yard. Convert from cubic feet to cubic yards:

> 528 cubic feet ÷ 27 = 19.5 cubic yards

Gravel to Greenhouse Ratio. We have determined a rough ratio of gravel volume (primary thermal mass) to greenhouse volume.

> 1 cubic yard of gravel per 180.5 cubic feet of greenhouse space

Framing

We used standard construction methods to attach the greenhouse to the south end of our garage. The vertical studs in the glazed walls are 2 x 4s. The front glazed wall studs are 2 x 10s. The thicker, unglazed walls are made of 2 x 6s. All studs are 2 feet apart, called "2 feet on center." The roof pitches, front and back, are at 45 degrees, an angle that is close enough to the ideal solar incidence angle for solar gathering and easy for carpenters to work with. We installed a vapor barrier of strong plastic sheeting between the garage and the greenhouse.

Framing: (Clockwise from top left) The back garage wall, the west face of the greenhouse, the southwest corner of the greenhouse, and the south face from the inside. Notice that the south wall has a 4-foot knee wall. We put this in to make the soil accessible right up to the wall. Also notice the extra black vapor barrier between the greenhouse and the garage.

We built the packing area in the garage, putting in a pass-through door, which we call the "Door into Summer," after the Robert Heinlein story about a cat that insisted that if it were just let out of the right door in winter it would find the door into summer.

Insulation

An important factor in any construction in the North Country is R-value. This is a measure of how quickly heat leaks through a building material—the higher the R number, the slower the heat transfer—and the warmer the structure.

In the unglazed walls and ceiling we used insulation with R-values around 20, using two layers of particle-board sheathing, siding, vapor barrier, pink fiberglass batting, and plastic vapor barrier. We used standard 4-inch fiberglass pink batting on side walls. The back wall and north roof have 6-inch bats. The foundation is insulated with 3-inch blue foam. The glazed walls were covered in twin-wall polycarbonate, which has an R-value of about 2, which means that it's a so-so insulator. (There is a heavier triple-wall polycarbonate, but we considered initial expense, its higher R-value, and lower light transmission and decided that twin-wall was sufficient.)

"Door into Summer"

With sunlight alone, the greenhouse temperature runs about 20 to 30 degrees F (–1 to –7 degrees C) above the outside temperature even on cloudy days. With a well-warmed gravel bed, it has no trouble staying at 50 degrees F (10 degrees C) for two or three days after a sunny day.

Light Conservation

The glazed walls are covered in twin-wall polycarbonate, which allows most light to enter while conserving heat and creating the light–heat transfer barrier. This isn't just about keeping heat in: When light passes through a clear substance into a room, then bounces around, its wavelength changes, meaning that it becomes heat. Polycarbonate is very good at keeping this "changed light" from escaping.

Polycarbonate also diffuses the light, which is beneficial for plants. The diffused light coming through the polycarbonate does not cast direct sunlight, and therefore the greenhouse lacks the very bright and very shady areas found in a space that uses a clear covering like untreated glass. It gives all the plants better access to the available light throughout the day.

We also painted the walls and rafters white to increase the light level. In any such project there's a balance between using sunlight as light and turning it into heat. Dark surfaces change more light to heat, while lighter surfaces bounce it around more. In the wintertime, duration of light is actually harder to come by than heat, so we chose light over heat.

Ventilation and Supplemental Heat

Any greenhouse needs air circulation—healthy plants require it—so fans are necessary. In a passive solar winter greenhouse this is even more important, because areas near the walls cool more quickly, so heat needs to be redistributed. For this we have a good-quality cage fan that we can move from place to place or hang on the wall.

Fan and Heater: The east wall, with the north heat-gathering pipe tying in to the east PVC vertical, the outlet vent fan, and the small propane heater, which rarely runs because it is only for backup heat.

Vents are also vital in a passive solar winter greenhouse: they exchange air in a controlled way, remove excess moisture, and cool the air, which can indeed get too warm. Our greenhouse has two vents: a powerful outlet fan on the east wall over the propane heater and a passive louver on the west wall over the sink. Both have hatches on the outside that can be firmly latched shut in the nastiest weather. We open the door between the packing area and the greenhouse when more ventilation is needed.

Two lines of black stovepipe run east-west on either side of the roof ridge. These capture the heat that accumulates there. The pipes are attached to the white vertical pipes that tie into the underground drain

Heat-Gathering Pipe and Blower: 1. The south heat-gathering pipe. 2. The tie-in to the vertical pipe on the west end. The blower is inside the vertical pipe (see arrow). The control thermostat is the oblong box just above it. (You can see the north heat pipes on the opposite page between the heater and the vent fan).

pipe near the sink and the propane heater. At the joint we inserted off-the-shelf, forced-air, heat duct booster blowers, which we tied into thermostats. When the greenhouse gets warm enough, about 65 degrees F (18 degree C), the blowers force the warm air into the gravel bed in order to store the heat.

Layout

We mulled over several raised bed layouts, selecting this one because it provides the most accessible growing space. We considered headroom, how far over a bed the gardener could reach, and walking space. The "keyhole" design we settled on allows the best accessibility with the least space taken up in paths. It also allows for dense hanging planters that can be pushed aside to access beds.

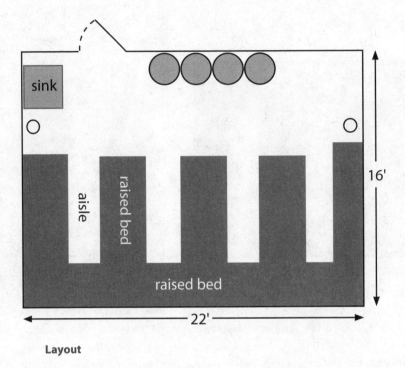

Layout

Hanging Planters

We use hanging trays to greatly expand our growing space. These are simply standard 10-foot plastic rain gutters cut into thirds, making three 3-foot-long, 4-inch-wide planters. The ends are finished with standard end caps. We drill seven evenly spaced 1/8-inch holes along the middle of each gutter's bottom for water drainage. The trays are suspended from the ceiling with ordinary rope—clothesline cord will do, but other types may be cheaper and more durable. Be imaginative.

How the Passive Solar Winter Greenhouse Works

The Garden Goddess greenhouse is a *passive solar* greenhouse because, unlike conventional greenhouses, it uses sunlight to provide nearly all of

Hanging Planters

the energy used to grow a season's worth of produce. A purist might call it "active" because of the two small blower fans in the heat capture system, but we consider that energy input to be small enough to be insignificant.

The design elements we described above work together in this passive solar system: The directed *glazing* gathers only useful light and minimizes heat loss. The *insulation* keeps in the heat generated by the sunlight. The *heat capture system* makes the best use of that heat. The *in-ground heating* puts the heat where it will do the most good. And the *thermal mass* stores the heat that's been generated and captured.

These five elements reinforce each other in a holistic system. We've seen other greenhouses that used some elements but not all of them—but they failed. They either froze up or took far too much fuel to heat.

FYI for DIY (and Collaborating with Your Contractors)

This manual should give you a solid start on what is an unusual but rewarding project. Any moderately skilled do-it-yourself builder can tackle it. It can also be a good community project, undertaken by a school, church, or neighborhood group with oversight by a knowledgeable person. You could organize a greenhouse raising party, like an old-fashioned barn raising!

You can show this manual to your contractors to assure them that nothing in it is radically new or hard to accomplish, just put together in an unusual configuration. All the construction techniques are standard for modern house building. Requirements for sewer, water, electricity, and heat are straightforward. Your contractors will probably have helpful ideas for improving the structure and systems, but be sure that they understand what you're actually trying to do. This manual can also serve as evidence that at least one successful passive solar winter greenhouse exists. Who knows—building your greenhouse may become as exciting to your contractors as it is to you!

Design Requirements Summary

Make substitutions, and improvise freely in your plan, as long as you know what you're doing! Just remember these requirements for a successful passive solar winter greenhouse:

- Use glazing only where the sun directly shines in.
- Store all available solar heat.
- Insulate the foundation, which must extend below the frost line.
- Use passive in-floor heating with access to supplemental heat.
- Heavily insulate all nonglazed walls.
- Make the interiors highly reflective.
- Control airflow.

Working Drawings

Plan View: Superstructure

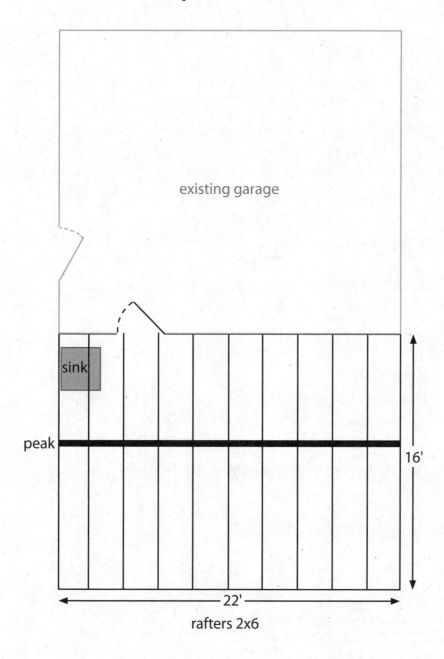

existing garage

sink

peak

16'

22'

rafters 2x6

North and South Elevations

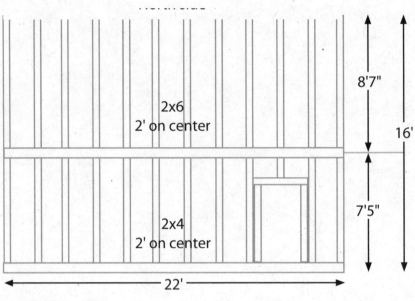

North Elevation

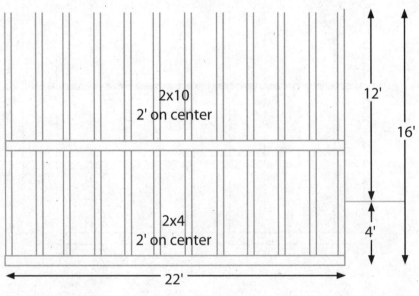

South Elevation

West Elevation

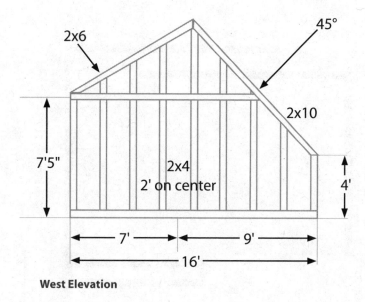

West Elevation

The East elevation is a mirror image.

Heat Pipes

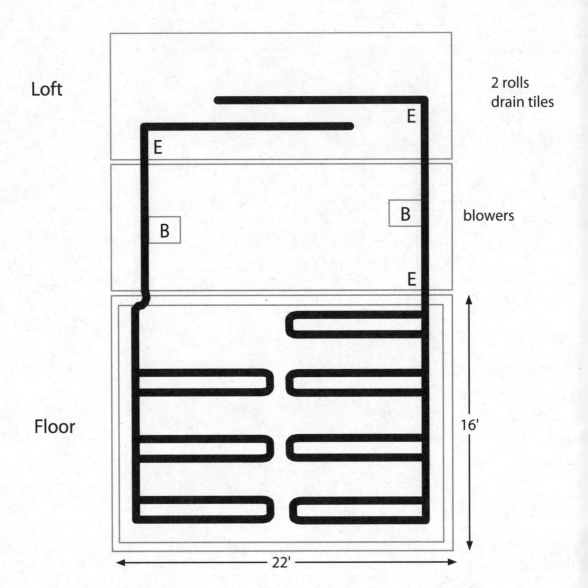

Loft

2 rolls
drain tiles

E

E

B

B

blowers

E

Floor

16'

22'

Garage Roof Tie-Ins

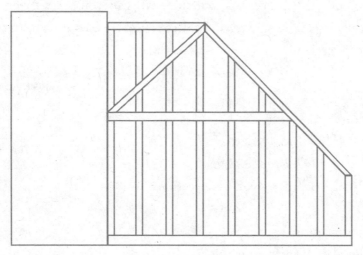

Top

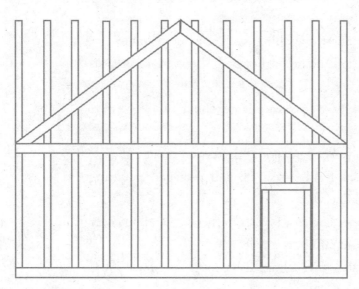

Bottom

Basic Materials Lists

These are not all-inclusive lists, but they will give you an overview of the scope of the project. Be sure to adjust these numbers according to the specifics of your project.

ITEMS   NOTES

For the Frame
___ 2 x 4s: **30** 8-foot lengths
___ 2 x 10s: **24** 12-foot lengths
___ Particle board: **6** 4 x 8–foot sections

For the Foundation
___ Standard 8-inch cinder block: enough for
76 linear feet, 4 feet high
___ Cedar plank (for the sill): also 76 linear feet,
18 inches high

For Sheathing
___ Standard sheathing, such as exterior-grade
particle board or plywood: about **14** 4 x 8–foot
sheets
___ Twin-wall polycarbonate: about **12** 4 x 12–foot
sheets
___ Polycarbonate mounting tracks: about **100**
linear feet between H- and Y-shaped mounting
channels
___ Vapor barrier (Tyvek or other): enough to cover
about 700 square feet
___ Outer vinyl siding: also enough for about 700
square feet

ITEMS	NOTES

For Insulation

___ Blue foam/expanded polystyrene: about **10** 4 x 8–foot sheets, 3-inch thickness

___ Fiberglass batting: about **24** 8-foot bats, 4-inch thickness, and about **35** 8-foot bats, 6-inch thickness

For the Roof

___ Tar paper: about **180** square feet (about "2 squares" by standard roofing-contractor measure)

___ Shingles: about **180** square feet (about "2 squares")

___ Flashing: depends on how you tie in to existing structures

For the Solar Heat System

___ Drain tile: about **160** feet of 4-inch perforated, and **14** 4-inch T-joints

___ PVC piping: **8** feet of 4-inch pipe

___ Duct blowers: **2** standard forced-air heat in-duct boosters

___ Stovepipe: **60** linear feet and **4** adjustable elbows

For Air Flow

___ Vent: **1** louvered, about 18 inches square

___ Vent fan: **1** louvered, about 18 inches square

___ Circulation fans: **2 or 3** 18-inch box fans, industrial preferred

ITEMS	NOTES

For Electricity

___ This list will depend very much on your particular layout.

___ Breaker box

___ Outlet boxes: **6** insulated

___ Cable

For Water

___ Sink: **1** laundry-type deep sink

___ PVC/PEX piping

___ Hose: **50** feet of black garden hose

___ Nozzle: long adjustable wand

Odds and Ends

___ Polycarbonate screws with washers and gaskets

___ Roofing nails

___ Framing nails

___ Duct tape

___ Greenhouse thermostats: at least **2**

___ Backup heater

___ Shop lights: **2 or 3**

GROWING THE PRODUCE

Autumn:
As Good a Time as Any
to Fall in Love with Soil

While the world outdoors prepares for a season of slumber, the greenhouse bustles with activity in autumn. The raised beds and planters are readied for their winter season of growth and harvest. Soils must be mixed, seeds sown, transplants placed in raised beds, and growing supplies stocked up and stored away for the winter. There's not a lot we can do about the dwindling daylight, but there's much to do to ensure that the season's vegetables have all the nutrients they need to do their best.

Solarization

Early in September, the raised growing beds in the greenhouse are prepped for production after a summer of solarization. What's that? Well, the greenhouse is a contained space, so we don't have nature's methods of checks and balances working for us in there. Solarization is the attempt to control the possible carryover of pests or pathogens from one year to the next by heating the soil. It is a step back from the extreme of sterilizing the soil.

solarization
"cooking" a greenhouse, using sunlight, to kill plant and animal pests

Perhaps you've read about using sterilized soil when starting seeds to avoid damping off—that nasty fungus problem that causes baby plants to rot at the soil level. Sterilized soil in combination with well-cleaned containers helps reduce the likelihood of a fungus attack. The usual recommendation is to heat soil up to 180 degrees and keep it there for thirty minutes. By definition, however, sterilization kills everything in the soil. But good guys are in there too, and for the raised beds of a greenhouse, it's preferable to strike that delicate balance between killing as many bad guys as possible and keeping some beneficial microbes around to do their thing. While my research showed some variance in what's considered ideal for solarizing soil, it seems to settle around 120 degrees for four weeks. That's pretty easy to accomplish in a passive solar greenhouse in the summertime. All you've got to do is close it up for a few weeks and let it bake. This kills off all the big bugs, of course, too.

We learned the hard way that you have to remember to get all the equipment out of the greenhouse that can't take that temperature before the process begins. Thermometers burst. The solar hydraulics of our automatic vent openers bit it, too. The heat also killed a clock. Lesson learned.

When the solarization is taken care of, I reintroduce good bacteria and microbes by adding compost and/or vermicompost. More on that, later, but I just wanted to drop that mention now to make you aware that it is possible to keep the soil healthy by adding back the heroes and helpers.

Soil

If you haven't already spent some time getting to know the fascinating world of soil life, I think you've inadvertently cheated yourself as a gardener or farmer. But that means you're in for a treat. Soil is an amazing, rich universe of bacteria, fungi, invertebrates, vertebrates, humus, and minerals all mixed up together. At the tiniest level of root growth, your plants interact with all the living and dead materials in the environment around them. Some fungi even grow around and in plant roots in a mutually beneficial arrangement, feeding each other. These mycorrhizal fungi help plant roots draw nutrients and moisture from the soil. The mycorrhizal fungi produce chemicals that break down nutrients for the roots to absorb. In return, the fungi gain access to the sugars produced by the plant's photosynthesis.

Something in me always knew there were minute miracles happening down there in the soil, but I didn't realize just how complex it all was. Soil biology is the science that food producers need to master the art of nurturing plants. One simple thing every gardener knows is that you can't grow squat in crappy soil. And most folks know that the best way to turn crappy soil into productive soil is to add lots of compost. Good compost is chock full of all those little mycorrhizae that hook up with the tiny threads of plant roots to funnel the goodness of the compost (the real black gold) into the foods we eat. It is just so cool. And that's only one story of what's going on in the soil beneath our feet. It's a zoo down there!

mycorrhizal fungi

a fungus that has a mutually beneficial association with plant roots

My fascination with the amazing and complex world of soil is the biggest reason I chose not to grow my greenhouse crops in a hydroponics system. I like to play in the dirt. With hydroponics, the roots are suspended in a water bath or are grown in a soilless medium that gets washed in a nutrient bath. For me, this is too much chemistry and not enough biology. And you end up with waste water in a hydroponics system, and I really don't want to have to deal with all that. The bottom line is that I just dig soil magic. Good rich soil smells of life and feels good in your hands. I love the tactile part of the gardening experience. I've learned to read soil with my hands like a baker reads dough.

In the greenhouse, I am the goddess of the soil. Higher powers created the stuff, but I combine those materials thoughtfully to suit the enclosed environment where the plants will grow. That's an important part of green-house production to remember—you are *not* in the summer garden. A green-house is an enclosed, intensive growing space. The rules that apply outdoors don't fly here. You'll read examples of that many times in this manual.

So you get the idea—in the early fall, I play, play, play in the dirt. This is particularly fun on a cool rainy autumn day, when most gardeners are forced to stay inside. I'm inside alright, but I'm whistling a happy tune, up to my elbows mixing soil components, enjoying their fragrance, and relaxing to the soothing sound of raindrops falling on the south wall of the greenhouse. It's divine.

None of the magic of good rich earth is lost on me now that I'm a student of soil. There's no graduation ceremony from that school because nature is all about complexity on levels that ensure I'll never run out of things to learn. But that's okay. I can be like the fungi, breaking down the lessons into digestible amounts and sharing the results with others. Soil savvy grows just like plants. The fun part is that you can really see the results as you apply your increasing knowledge. And you don't have to be a soil nut for that to happen. The basics will get you there.

One last thing I'd like to add is a recommendation about soil testing. It's wise to test garden soils outdoors simply because there's no point in putting nutrients into your soil that it doesn't need. That kind of overkill contributes to fertilizer runoff, which creates serious problems in our watersheds. And testing is even more important in the greenhouse.

You have to be savvy to get the mix of soils right for optimal growth. Start out with a balanced soil mix and replenish it at the beginning of the growing season and also after a crop is harvested. This seems like a system that wouldn't really need a test to ensure that no nutrients are absent or overly abundant. But you know, it just can't hurt to find out for sure.

That's why I test my greenhouse soil. This isn't necessary for the planters, because after a crop is harvested several times, the whole thing gets dumped in the compost and the planter is refreshed with new soil. No worries there. Any leftover nutrients benefit the growing beds outdoors in the form of finished compost. But in the raised beds of the winter greenhouse, an abundance of any nutrient could harm your plants. This is because when the days are short, plants have different needs than they do in high summer outdoors. Too much nitrogen, for example, can cause high levels of nitrates in plants that don't respirate, or "breath" much during long stretches of cloudy, cool weather.

It's not that big of a deal to do a soil test. Buy a basic kit, follow the instructions, and note the results in your greenhouse journal (see page 60). It's also a good idea to check for an accumulation of salts and minerals after the first couple of years of production, because a buildup of salts is going to affect the productivity of your plants. I have not yet had this problem in my greenhouse, but I do check for it.

Preparing Soils for Your Greenhouse

The soil mix you use in raised beds and planters needs to provide the root systems of your plants with the nutrients, moisture, oxygen, and support they need to be as productive as possible. There's more than one way to make this happen. I've broken down the basic ingredients of the soil mixes that I use, and I've also included some other possibilities for you to consider so you can do your own testing and see what works best for you.

Soil Amendments

Here is a brief list of the soil amendments I use in the greenhouse—both in the raised beds and in the hanging planters. (You can find the recipes on page 44–45.)

Peat Moss

Peat is the result of an ecology in which decomposition does not keep up with new plant growth. Peat is a mat of partially decomposed vegetation beneath the growing plants in a bog that—thanks to anaerobic, acidic conditions—builds up over time. When you use this product in your soil mix, you need to add some lime to offset the acidic nature of peat. Peat doesn't have a lot of nutrients, but it provides humus, retains moisture, and creates the air space that roots need to be healthy and happy.

There is some controversy out there regarding the use of peat because it is a finite resource and one that takes many years to replace after it has been harvested. In some regions of the world, pressed and dried peat is used as a fuel source, rapidly depleting supplies there. In the United States, some areas of peat production are sensitive wetlands, and the disturbance of those areas is troublesome to conservationists. Others argue that well-managed peat harvesting can be done in a way that allows areas to recover from the harvest.

Some organic growers are experimenting with substitutes for peat. One of these is coir, a product made from shredded coconut husks. I am doing some experiments with it to see if it's a good peat substitute for me.

Peat moss   **Coir**

In this manual, though, I am being pragmatic. With experience, you will begin to experiment, as we have. But if you are a beginner and you need a bulk material that is easy to find and not too expensive, that would be peat. It comes in bales at most home and garden stores. Watch out for the cheap stuff. It can have sticks and stems in it that add to the weight but have to be sifted out of your soil mix. When you find a brand that gives you good quality peat, stick with it. I've bought cheap peat on sale, and it was no bargain in the end.

Vermiculite

This soil amendment is superheated rock that expands into a flaky material. It keeps the soil mix fluffy, helps retains moisture, and adds some trace micronu-trients. Roots need air space in the soil, and vermiculite helps with this.

Vermiculite

Perlite is an alternative to vermiculite. You've probably seen it in indoor plant potting soil mixes. It's those white chunks that turn green over time. Perlite is also superheated, expanded rock, but it does not absorb moisture, nor does it add anything in terms of soil nutrition. It's just there to take up space and turn that green color I find rather gross. Perlite also kicks up a great deal of dust when it is being incorporated into a soil mix. It's always best to use a breathing mask when you're mixing soil, but even with that, perlite is a real dust devil.

I find vermiculite a far superior product—at least for now. I've talked to the owner of a local florist/nursery business who is interested in alternative soil components. He wants to use a local product to replace vermiculite, and he's trying wheat chaff. I plan to try out some of that stuff too, and I'll let you know what I think at our website, gardengoddessenterprises.com.

I don't want to get completely locked down with the materials I use for soil improvement. I just want to make sure the new stuff is better before I completely switch over. And again, vermiculite and perlite are not hard to find. The list of suppliers at the end of the book will help you locate your own source.

Compost

Compost is an important ingredient in the soil mix. It contains nutrients and organisms beneficial to your plants. I purchase bagged organic, locally produced, plant-based compost. Oh sure, I make compost out in the yard in summer. That stuff gets spread throughout my garden plots to enrich the earth there. People ask me why I don't use my own compost in the greenhouse as well. It's because I am leery of using that material in the greenhouse. It's not that the outdoor compost is bad—not for the outdoors, anyway. Those wide open spaces are designed to handle whatever's been cooking in the compost pile. But I am protective of the soil in my greenhouse.

Compost

Again, the greenhouse is a microcosm unto itself. It lacks the great outdoors' ability to use rain, wind, climate, and a vastly diverse ecosystem to keep good guys and bad guys in balance. All of that work is up to me in the greenhouse soil, so I need to control what goes in there. If my outdoor compost ends up bringing a disease, soil pest, or unwanted fungus into the greenhouse, the natural systems to control it are absent. So it may seem a little picky, but I err on the side of caution with the greenhouse soils. I don't want to get stuck with a problem that is difficult and costly to fix or that ruins a winter's worth of vegetables.

But I am experimenting in this area as well. I have talked to other farmers who use vermicompost (worm castings). They claim that vermicompost is superior to plain compost because the nutrients in organic material that has passed through a worm's digestive system are more readily available to plants. I am now comparing transplants that grow in my usual soil mix with transplants that grow in a vermicompost/coir mix. I can't wait to find out who is the winner. So far, I have noted that the seedlings in a vermicompost-enriched soil germinated at about the same time as those in my standard soil mix, but grew faster during the first few weeks. Once transplanted, they all seemed to even out in size. I do think that vermicompost is a great way to incorporate beneficial organisms into the soil, and I intend to continue my exploration of its benefits and the best ways to make use of it both in the greenhouse and in my outdoor garden plots. We've started our own vermi-compost project to see what the little red wiggler worms can teach us about soil improvement.

Curiosity keeps me working like a little mad soil scientist in my green-house school. This book is a snapshot of what I know so far, so you can take it from there. Won't it be fun to find out how the picture changes over time? I just never get tired of this stuff—it's my lifelong soil school.

Organic Fertilizer

The final touch in the soil mix for the raised beds is organic fertilizer. Generally, I use equal parts greensand, rock phosphate, and blood meal. Each of these components has its own job for boosting plant production.

Greensand (glauconite) is a mineral deposit formed on ancient ocean floors. It supplies potassium, which helps in plant metabolism (building protein, photo-synthesizing, fighting off disease). Greensand also contains silica, iron oxide, magnesia, lime, phosphoric acid, and twenty-two trace minerals. It looks and feels, as you might imagine, like very fine sand of a sage green color.

Rock phosphate provides phosphorus. Phosphorus is essential to photosyn-thesis—the amazing transformation of solar energy into chemical energy. It aids in the formation of oils, sugars, and starches; helps plants withstand stress; and encourages root growth.

Blood meal is the nitrogen component of the fertilizer mix. Nitrogen helps plants with vigorous growth and is a vital part of photosynthesis.

Organic fertilizer components

You'll want to buy these fertilizer components in bulk because (a) it's cheaper and (b) you don't want to run out in the middle of winter. I found this out the hard way, but the fates were kind to me. I contacted a couple of local garden/nursery business owners to see if they happened to have any of these materials on hand. I asked if I could buy from them and piggyback on their future orders. I found I got better-quality product at a comparable price. It's good to know folks in the business. And they're always willing to talk greenhouse production with me. Gotta love that.

Putting It All Together

I use the raised beds in the greenhouse for larger crops that have bigger root systems, like broccoli, chard, Chinese cabbage, and kale. Some of these crops grow in the greenhouse all winter. Others are harvested out, and new plants take their place. Greens in the planters have a shorter life than the slower-growing crops in raised beds, and when the greens are done, they and their soil are removed from the greenhouse, to be replaced with new soil and new seed. I use a slightly different soil mix for raised beds and planters, but the ingredients are the same.

My soil mixes are similar to the one used by Eliot Coleman, which he describes in his book *Four-Season Harvest*. I combine the ingredients in an old wheelbarrow that resides in the greenhouse for this purpose. I use a hoe to do the mixing and sometimes my hands to break up clumps of compost or peat—you want to make sure everything gets well mixed. Besides, it feels good to get your hands in there. The texture and smell of the finished product seem to beckon seeds and water to come join the party. I know a farmer who uses a spinning composter barrel to mix her soil ingredients, and that works great too.

Carol's Garden Goddess Soil Mix Recipes

Note: My measuring bucket is a 1-gallon ice cream container; 1 bucket = 1 gallon.

Raised Beds

When I first created the raised beds in my greenhouse, I combined equal parts black dirt, peat, and compost to create beds that were 6 inches above floor level. Each fall I add more soil mix to that, and now I have 8-inch beds—so some settling and removal of soil happens along the way.

> 7 buckets of peat
> 1 cup garden lime
> 2 cups bloodmeal
> 2 cups green sand
> 2 cups rock phosphate
> 1 40-lb. bag of compost

Measure peat into large container. Add lime to peat, and mix well. Combine bloodmeal, greensand, and rock phosphate. Add this mixture to peat, and mix well. Add compost to the other ingredients, and mix well. I spread this final mixture on top of my raised beds at a depth of 2 inches and incorporate loosely with the hoe. After I smooth the surface of the raised bed with a rake, it is ready for transplants.

Planters and Soil Blocks

3 buckets of peat
½ cup lime
2 buckets of vermiculite
1 cup greensand
1 cup rock phosphate
1 cup blood meal*
3 buckets of compost

Mix the peat and lime together. Add the vermiculite, greensand, rock phosphate, and blood meal, and mix together. Add compost, and mix again.

Note: During the middle of winter I cut the amount of blood meal (nitrogen) in half to lessen the risk of nitrate buildup in the baby greens. More on this later.

As you can see, there is nothing terribly complicated about putting together the components of the soil mixes you need to successfully grow crops in your passive solar winter greenhouse. Just make sure you stock up on what you need to make it through a winter. I estimate that I go through twenty-five 40-pound bags of compost, five large bales of peat, and two large bags of vermiculite each winter. One thing is for sure: it's way better to end up with some extra supplies in spring than to have to scramble around in February trying to find somebody who will sell you a bale of peat.

Raised Beds: Planting in Early Fall

If you read the section on building a passive solar winter greenhouse first, you know that we excavated 4 feet down when we built the Garden Goddess greenhouse, putting down a layer of rock that has drainage tile in it (for heat storage) and a layer of black dirt on top of that.

I did some research to find out what sort of configuration of raised beds would give me the most square footage of growing space. The one I chose starts with a raised bed that runs along the outer edge of the greenhouse along the east, south, and west walls—like this:

Alley, raised beds, and sink
(facing west)

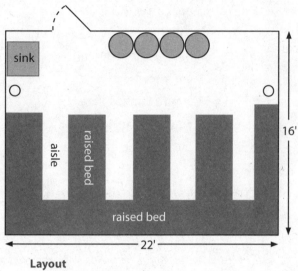

Layout
(footprint; north wall at top)

The bottom line shown here represents the raised bed that runs along the south wall of the greenhouse. In addition to the two raised beds on the east and west sides of the greenhouse, I also constructed three raised beds in the middle, with access aisles between them. The finished layout looks like this:

This configuration allows me to access the south raised bed, as well as all the hanging planters, from the aisles. The back (north) wall of the greenhouse has several plastic drums of water with a board on top of them. This is where I propagate seedlings. The rest of the back wall gives me space to store mixed soil, my wheelbarrow, and my tools. There is an alley between the barrels and the ends of the raised beds.

It takes a lot more raw materials than you might think to make 6-inch-high beds. I knew that I would continue adding soil components to the beds over the years so they would gradually get deeper with time (and they have). In addition to the basic start up soil, I spread a ½-inch layer of my basic planter soil mix on top, to give the beds the benefit of the mineral fertilizers and some vermiculite. I worked this into the surface of the bed with my rake, then smoothed and shaped the final beds with my rake and flat hoe. Boy, was it satisfying to look over that project when it was done. All those pretty beds of soil full of promise seemed to whisper, "Plant me, plant me—let's get this show on the road!"

The great thing about these raised beds is that nobody ever tromps on them and rain downpours never touch them. These beds are not subject to the usual problems that lead to soil compaction. They remain fluffy and loose. Each fall, I add another round of 1–2 inches of basic soil mix to the surface, loosely work it in, smooth, and plant. When I remove the root systems of harvested plants, I shake out as much of the soil around those roots as I can. After four years of this, I have some pretty dynamite soil in those raised beds, which are now 8 inches high. I have never had any kind of soil problem in the greenhouse (knock on wood). I hope I can continue to perfect those growing beds, because they churn out an impressive amount of produce every winter.

Popular Space Hog, Broccoli: A Raised-Bed Case Study

By far, the most popular raised-bed crop among the Garden Goddess Produce shareholders is broccoli. The great thing about greenhouse broccoli is that there are no evil white cabbage butterflies fluttering about, laying their dastardly eggs all over the helpless plants. There are no fat green caterpillars in my greenhouse broccoli. Ever. No, those beautiful plants look like a line of perfect models strutting down the isle in all their glorious jade perfection. And broccoli grown in the winter greenhouse does not taste as strong as summer broccoli. It's tender and sweet.

The trick with the winter broccoli crop is to get it started early in the greenhouse. This is indeed a trick because our nearly perfect winter greenhouse has one little flaw: its heat-retaining design works too well in the fall. In the depths of winter, every opening for vents or fans presents the problem of leaking cold air. So we opted for the most minimal venting system. It works great in January, but in early autumn it can get hot in there. Broccoli is not happy about hot.

I start the seedlings for the first greenhouse crops in early September. If we get warm or hot days, the flats of plants can be taken outdoors and covered with pieces of old row cover to protect them from nasty bugs. This is a little risky because it could give an outdoor pest the opportunity to make its way into the greenhouse, but the broccoli needs an early start, getting planted as things are cooling down, so the plants put on some size before the day length really starts to shorten. If broccoli gets too hot in the greenhouse, it won't develop properly, and the stress will make it more prone to disease and pest problems.

When the broccoli babies have their second set of true leaves, they go into the waiting soil beds. They are joined by other mainstays of the Garden Goddess greenhouse floor, such as pac choi and Chinese cabbage and chard. I'm always trying something new. Not all vegetables do well in the winter months, even ones you'd expect would perform well. This year, I'm trying a couple of varieties of kale to see how they do.

Seeding Plants for the Raised Beds: The Soil Block Way

I do not seed directly into the raised beds for two reasons. First, the plants germinate and begin their lives more vigorously if started on heat, like on a propagation mat. Second, every inch of growing space in the greenhouse is precious and must be used to its utmost efficiency, so I don't want bare spots where germination failed for one reason or another. I also don't want the hassle of thinning in a crowded, intensive growing space. Thus, I need to start the plants that will end up in the raised beds in a smaller space, like a standard propagation flat. If you haven't seen them, flats are typically 11 by 21 inches and a few inches deep, and made of black plastic. Some have drainage holes; some are solid. I like the solid ones, myself, because of how I use them.

I grow my transplants in a soil block system. It uses the same soil mix as the planters but in a different way. The planters get filled with dry soil mix and are well moistened after seeding. To make blocks, the soil mix must first be made very moist, because the blocks are 2-x-2-inch compressed squares of soil mix. The moisture is needed to hold them together during the compression process. I know it sounds a little weird at first, but stick with me here and I'll explain, because I think you need to know about this method of raising healthy transplants.

The first thing you need is a soil block maker, a rectangular metal tool with a handle on top (information about where to find tools is in the last chapter, Resources: Seeds, Bugs, Equipment). You push the thing down into moist soil mix to pack the four square openings full of soil. The handle at the top has a release mechanism that you squeeze when you position the soil block maker onto a plastic flat. Squeeze, and then lift, and . . . voilà! The first four blocks. There is even a nipple in the top wall of the block so when the soil block is released, it already has a hole ready to receive seed. A standard flat can fit fifty blocks in it, which (as my mentor pointed out as she showed me how to do this) makes for easier record keeping: 2 flats = 100 transplants.

To fit fifty soil blocks into a standard flat, you just continue making rows of four, one after the other down the length of the flat. This gives you ten rows. Four soil blocks do not completely fill the width of a flat, so when you have the first forty blocks done, turn the blocker sideways on that empty side, and you can fit two more rounds of four blocks in there. Ah, now you have forty-eight. To get the last two in there, make a round of blocks that you release in the soil mix container. Pick up

Soil block maker

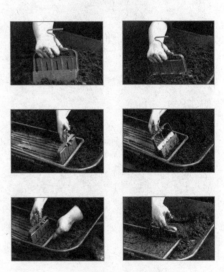

Making soil blocks

Soil block . . . made

one gently and place it in the remaining space in the flat. Do this one more time, and you've got your fifty. If it's no big deal to you whether you have forty or fifty soil blocks in your flat, just do what you want to do, and the seedlings will be fine. The blocks will hold together even if they are not cozy as can be.

Soil blocks and the tool of their maker

Rather than watering the soil blocks from above, you water them from the side. I gently pull the side of a flat outward and use a plastic watering can to pour water into the flat. The blocks soak up a lot of water. Give them some time to absorb what you've given them to see if it's enough or if they need more. You don't want them sitting in water or getting dried out.

I admit that it takes a little time to get the knack of using the soil block maker. I won't claim that it's completely simple, because I ruined my share of early attempts before mastering it. Have some patience with yourself and the tool. The right amount of moisture in the soil mix is key to your success. If the soil mix is too wet, the soupy mix is apt to slide right out of the blocker before you're ready to release it. Not enough moisture, and the blocks turn out crumbly. When I add water to soil mix, I go by feel. I squeeze a handful, and if it releases some water, it's likely good to go. Just remember that if you oversaturate the mix, you may have to figure out a way to drain out that excess water (use for watering other plants, of course). Or save some dry soil mix on the side so you can add it in to absorb an overabundance of moisture.

When your soil blocks are made, you simple place a few seeds in the hole of each block and then crumble in some additional soil mix. The amount of coverage depends on what plant you are growing, so check the seed package. The general rule is twice as much soil as the width of the seed. Press the soil down lightly with your finger so the seed has good contact with the soil around it and is better able to absorb the moisture that will trigger it to germinate.

You also need to mark your flats so you know what you planted and when. This is especially important if you are growing more than one kind of plant in a flat. I make markers out of old cottage cheese and yogurt containers. I trim off the top rim and cut off the bottom, then cut strips from the remaining plastic. I use a fine-tipped permanent marker to write down what I planted and when. I put these in flats and in my planters. I also write that information in my greenhouse journal.

When the seeds have germinated, you'll most likely have more than you need. After a few days of growth, determine which one in each block is the healthiest. Grab a scissors (one that you will keep in the greenhouse hereafter), and carefully snip the other seedlings at soil level. Tugging them out of the soil risks disturbing the one you want to keep. If you feel bad for the seedlings that didn't cut the mustard and must be culled, then make good use of them and eat them. They're really quite tasty, no matter what you're growing. Add them to a salad, sprinkle them on soup, or just munch as you work.

So why go to all the fuss and bother of soil blocks? I have two words for you: air and space. There is room between the blocks for air, which acts as a barrier to growing roots. The roots don't want to be exposed to daylight, so they don't try to push outward at the edges of the blocks. This helps prevent the common problem you've seen when you buy flower transplants in springtime at stores. You get them home, pop them out of their little cells, and find a wall of outer roots that grew to the cell's edges and started to circle around it. This kind of root-bound transplant is easily stressed because its roots are so exposed to dryness and heat. A young plant with roots poised and ready to shoot out into its transplanted soil home gets over transplant shock much more quickly than a plant with a mat of exposed roots that must be teased loose to function properly.

This leads us to our second word: space. A transplant in a soil block has well over three times as much soil to grow in as one in a typical transplant cell. I also like the fact that soil blocks reduce plastic use because all that's required to contain them is a typical sturdy, reusable transplant flat, not the flimsy cell containers that rarely last for more than one or two uses.

So think about the soil block maker. It may not be the thing for you, whatever your reasons, but I'm really sold on it. If you become a big fan, there are even larger soil block makers that crank out a whole flat's worth of blocks at a time. I give mine a good rinsing after it has filled a bunch of flats for me, because I want it to last a good, long time. Because of the basic care I give it, my block maker still looks brand new even after four years of hard work. I love that in a tool.

Planters: Planting in Mid-Fall

As September moves into October, the greenhouse wakens from its summer dormancy. The raised beds are filled with eager young plants. Outdoors, many crops are gathered and stored in the root cellar for the long months to come. And most exciting of all, the workhorses of the greenhouse are coming into full production.

At Garden Goddess Produce, we grow over twenty varieties of winter greens, using lengths of plastic rain gutter as planters (see page 23). The planters hang in tiers of three from nylon rope harnesses that run the length of the greenhouse. Two rows of these planter tiers fill the greenhouse's vertical space with a plethora of tasty salad greens. The planters are easy to care for and use. They produce all our baby greens, which are the most favored crops we grow. In fact, the winter greens sell our shares, because folks find them to be the best they've ever tasted.

One of the many good things about this method of growing greens is that it makes harvest much easier for you, the farmer—no tedious bending and cutting of bug-chewed, weather-beaten greens splashed with mud like you get from the great outdoors. The passive solar winter greenhouse is a protected and sunny but cool slice of heaven for greens. The planters are densely seeded and watered by hand, so mud splash is not a problem. It takes three to five cuttings of a crop before the soil is spent and must be replaced.

As you might guess, I have a lot of these planters neatly stacked away during the summer. When it's time to gear up for the next winter season, the planters get a

Alley and raised beds

good scrubbing with bleach water (1 teaspoon of bleach per gallon of water) and a good rinse. I use a vegetable scrubber to clean off the plant material that has stuck to the sides of the planters. After soaking a while in water, the dead stuff comes right off. This reduces the chance of any disease carryover in the planters. And as long as I'm at it, I give all my plastic transplant flats the same treatment.

It would be heartbreaking to go through the process of mixing soil, seeding it, and watching young plants germinate only to have them all die in one day from damping off or some other disease. A ruined crop affects your harvest schedule and your sanity.

I can fit eight planters on the growing bench where I have two propagation mats. When it's time to do some seeding, I make up a batch of soil mix and fill eight cleaned planters to the top with the dry soil mix, lightly press down with my hands, and then water it thoroughly to moisten the mixture.

I wish you could watch me seed a planter so I wouldn't have to figure out how to explain the amount of seed needed. The amount varies according to the size of seed in a given planter. I do it by sight, like good cooks use herbs. I just know the right amount, and I don't have to think about it. But for you, I'll try.

Seeding a planter

It's hard to see dark seeds in dark soil mix, so I have chosen photos of recently germinated planters of greens. You can see approximately how many seeds were spread in a planter based on the number of seedlings that emerge. Of course, a few seeds are duds, so bear that in mind when looking at these photos.

It's important to remember that different varieties create different sized plants, even at the baby stage. You will need less seed for a thick planting of baby chard than you will for a thick planting of lettuces. Planting some varieties too thickly is counterproductive because a dense mass of greens doesn't dry out properly after watering and therefore can be prone to disease problems associated with too much moisture. Minutina and claytonia, for example, are likely victims. You may be in for a little trial and error to get the dense planting thing down, but I do note in the list of greens varieties (starting on page 94) some of their specific needs.

Seedling density

After seeds are sprinkled into the planters, they are followed by a thin covering of soil mix, which is also pressed down and then watered to make sure the seeds have good soil contact. When the planters are seeded and watered, I place them on the bench at the back of the greenhouse. (The bench is actually a large board on top of several large plastic drums of water.) On that table are two large propagation mats laid across its length. The mats look like big heating pads, and that's basically what they are. Propagation mats help seeds sprout more quickly because they warm the soil from below. Yes, all these crops like cool temps while they're growing, but they also respond well to warmer soil while they're sprouting. I put both the planters and the flats of soil blocks on the propagation mats.

Heating bench and shelves

Once seedlings have sprouted and start their first true set of leaves, they come off the heat. If there's room in the slings, the planters can go hang out with their leafier brethren. If not, they can reside for a week or two on metal shelving we installed above the propagation table. Since initial growth does slow down in midwinter, the shelving has given me the advantage of having more planters started up. I know this much: there's no such thing as too many salad greens.

Young planters

The planters require daily watering, especially when the temps get above 70 degrees F (21 degrees C) in the greenhouse. During a stretch of cold, cloudy weather, you can probably skip a watering day or two. On a hot autumn or spring day, you may need to water planters twice. I use the finger method to determine their water needs: I stick my finger into the soil. If the soil feels dry, it needs watering. If it feels moist in the morning, I still might give it a light watering if a sunny day is predicted and I know the greenhouse will really warm up. If the soil feels damp, then I skip the watering.

Greenhouse fields

I use an extendable watering wand to do my greenhouse watering. This allows me to get at all the raised bed plants as well as the planters. I check the raised beds in the same way I check planters to find out if watering is required. Watering is my time to look over the whole greenhouse for signs of problems that need attention. If the ends of the planters closest to the walls of the greenhouse have dried-up or frost-damaged leaves, I know I need to tweak my fan system to improve air circulation to those planters. They're getting hot or cold spots, which kill little plants. I also check leaves for signs of pest damage or fungus problems.

Making this sort of daily checkup part of the regular routine in the greenhouse is critical. It helps prevent little problems from becoming disasters. It's also a pleasant way to spend a little time with the plants. I find it calming in the morning and evening to take a few minutes to move around in the greenhouse to observe what's going on. It's hard to call something "work" when it's so enjoyable.

Harvesting: Late Fall

In November, the greenhouse harvest kicks in as the outdoor harvest winds down. When there's time, final chores get done outdoors. But focus definitely shifts to greenhouse activity, and it's a lot more fun working there than it is digging up the last of the garden leeks in a cold rain. I love leeks, and they are worth the occasional mud-wrestling contest, but the greenhouse is a kinder, gentler place in early winter.

When it is time to harvest a planter of greens, I rely on my trusty harvest knife. It is small and light and has a sharp, serrated blade. It is important to have a good sharp blade because a dull one will do the same thing to your greens that the dull blades on a lawn mower do to grass. The remaining plant stems get a raggedy cut and are more likely to brown at that tip point. Besides, if you have to saw away at your greens, it increases the likelihood that you will mistakenly tug a small plant completely out of the soil. That's because while your dominant hand is doing the cutting, the other is holding gently but firmly onto the leaves that will then be deposited into your harvest basket. Harvest is a chore if you're working with a dull knife, so learn how to use a sharpening stone to ensure that your blade always has a good edge on it. You can find sharpening stones in some seed catalogs (see page 113 for a list).

I do feel compelled, however, to issue a mild warning: sharp blades do not distinguish between tender greens and tender flesh. When I am in the greenhouse harvesting at night, the coolest time, it is usually about 40 to 50 degrees F (4 to 10 degrees C). My hands end up scrubbing root crops in cold water or rinsing harvest baskets in cold water. The air isn't warm enough to get the circulation in my fingers going all that well, so after a while, my hands and fingers can get somewhat numb to it all. That's not always a great combination with a finely honed, sharp object like a harvest knife.

Your best bet for safety is obvious, despite the fact that I sometimes ignore it: warm up your hands. It's also important not to let your mind drift outside your greenhouse and the task at hand while you're harvesting. It's easy to do, since there is a certain Zen rhythm that goes with cutting greens. I find myself thinking about what's going to get planted for the next month, or even sorting out the crop rotation outdoors for the coming season, and next thing I know, Ouch! I have to stop everything and head to the packing shed for a bandage so I don't bleed on the greens. Focused attention and good circulation in the outer extremities will leave you with a good harvest and minimal scarring.

Red sorrel, minutina (in flower), lettuces, chard (clockwise from upper left)

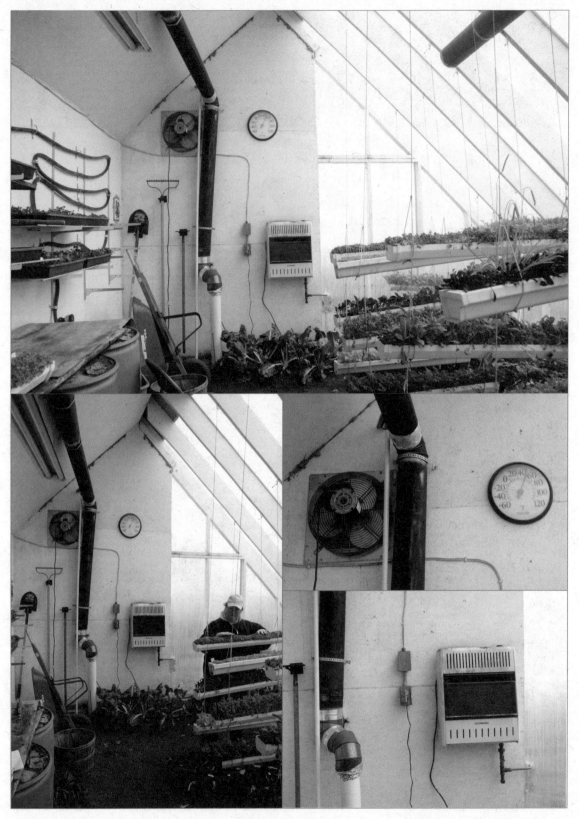

East wall

West wall

Osaka
purple
3/7/09

Red Giant
3/8/09

Tokyo
Bekana
3/7/09

Osaka
purple
3/7/09

Vitamin
Green
3/7/09

Vitamin
Green

Vitamin
Green

North wall and young planters

Soil ingredients, seeding planters, and dense seedlings

Hanging planters and raised beds

The Garden Goddess greenhouse, its proprietors, and some of its produce

Winter:
Growing Produce

Winter is the best time in the greenhouse. I don't have to worry about over-heating in there, and it is a mighty pleasant retreat on a sunny weekend. I've even invited friends over on occasion for an afternoon board game and some sun tea. Or I just grab a camp chair, put my feet up, and write in my greenhouse journal for a while. There's a radio if I feel like adding music to the scene, but most times I just enjoy the quiet, the aromatherapy, and the scenery.

Am I making this sound like nothing but an easy time? That's a bit mis-leading. There's plenty of work to do during winter production, but as far as I'm concerned, it's enjoyable, not difficult, and rewarding. You just need to know a few secrets to keep things humming in the greenhouse as the days grow shorter.

The Mystery That Is the Winter Planting Schedule (and How to Unravel It)

Here I'll discuss the tricks and triumphs I've accumulated in matters of winter planting. First off, forget what the seed catalogs tell you about germination time, harvest time, and all that. When you're growing in a winter greenhouse, it's a new day—literally. The combination of shorter days and cooler temps makes for a unique environment with its own timetable. I break that growing season into three mini-seasons: diminishment, solstice, and expansion. More on this concept in just a bit.

Your Greenhouse Journal

The best tool you have to ensure success with greenhouse production is your greenhouse journal. The more you record in it, the more valuable your journal will be to you in the future. Here are the things you will keep track of in your greenhouse journal.

What you plant and when.

In addition to noting the seed variety and planting date in my journal, I also make plastic ID tags for the planters and flats of transplants that indicate variety and date seeded. I make my ID tags out of strips I cut from empty yogurt or cottage cheese containers. I further appease my guilt over having such waste by using tags twice; writing on the bottom side the next year. I use a permanent marker so the information is easy to read later.

When a planting germinates
and when it is ready to move out to the growing area.

My planters move from the germination table to the planter harnesses when they have developed their first or second set of true leaves. The length of time this process takes depends mostly on the variety. You want to remember which ones are slow to get started so you can factor that into your plan of what to grow and when to plant the seed; then it will be ready when you want it.

How long it takes for the varieties
to come to harvest size and what they produce.

The time it takes for a planting to be ready for harvest varies quite a bit in the different greens we grow. In early season, most are ready to cut in four weeks, five tops. But they also vary on how soon they can be harvested again. Some are ready in only three weeks, and others can take six weeks before they are ready to be harvested again, especially during solstice season (see page 74 for more about solstice season). It's important to note variety performance in all the crops to refine selections you make next year.

How many cuttings you get before a planter is ready to be replaced.

At the risk of sounding redundant, the number of times greens are harvested varies both in variety and season. You will come to appreciate the greens that remain steadily productive during the solstice period. Most greens will give you three good-quality cuttings. Some give four. I usually taste greens before a fourth cutting to make sure they haven't become bitter or will soon bolt. Toughness and strong taste are not something I want in my salad mix.

Conditions in the greenhouse and outdoors.

This means noting temperature, day length, wind, cloud cover, precipitation, humidity, and soil temp. I note greenhouse air temp at the peak of the ceiling, at shoulder height, and at floor level. This can tell you a lot about air circulation.

Signs of disease or pest problems, plant growth, or stress.

Observation is a key part of keeping everybody in the winter greenhouse happy. I learned from an aphid infestation in my first year that it's a good idea to take a close look at plants and soil every day. If nothing else, it gives me something to write in the journal.

Observations and general impressions.

Such information can be more important than you know at the time you're writing it. Sometimes for me, it's even an idea I get for growing in my outdoor plots, something like, "Next year, I want to rotate the celeriac over by the garlic." "Lots of frost built up on polycarb overnight" may feel like a Post-It note to myself at the time, but when I look back months later, all those comments collectively give me important insights.

I admit that the greenhouse journal thing might seem like a bit much. There have been lots of times in my own life when I was told to keep a journal for this, that, or the other thing. I would diligently start it up and slack off within a month at most. For some reason, I stuck with this piece of my greenhouse chores, and it has proven to be a genuinely valuable tool. I also realize it simply doesn't take much time to maintain the journal. I keep a spiral notebook in the greenhouse and jot down daily data after I've checked over the plants, watered, or seeded planters.

A greenhouse journal is not meant to be a tome. Some days, the only thing in mine is temps, cloud cover, and what was planted. It doesn't matter. What matters is the accumulated information you gather over time. The greenhouse journal is also a great excuse on a cold weekend afternoon to grab my journal, beverage of choice, and favorite lawn chair to "work" in the greenhouse. When I sit down in that warm, sunny, humid growspace on a bright winter day, it's nice to be able to think of it as a valuable research collection activity. It's also one of my favorite perks of having this magical space.

The Planting Schedule (or Not)

So I've still not described my own specific planting schedule. Every gardening book I've ever read has these impressive, detailed graphs full of dates for planting and harvesting that make it look as if all you have to do is follow the plan and it will unfold flawlessly. Well, I realize now that those smarties are doing the same thing I have to do: generalize.

Many variables influence how well and how fast a plant grows. You might have a very cloudy December or one that offers ample sunshine. March can be kind or cruel when it comes to snowfall. One winter can be bitterly cold and windy, and the next will be blessedly mild. These variables—especially cloud cover—affect plant health and development in the greenhouse.

The grower has to dance a careful dance that responds to what's happening now and anticipates what will happen next. Each year increases the body of knowledge used to handle the unique conditions in a passive solar winter greenhouse, and that individualized information is golden.

I've learned small but important things about the crops I grow. Arugula, for instance, provides a plentiful first harvest and successive cuttings that are not nearly as flush as the first. If it suddenly gets hot, that second cutting may be your last, because the plants will bolt and turn bitter.

I've found that mizuna takes a little longer than other varieties to be ready for its first harvest. But the next harvest and the one after that come much sooner. And in the raised beds, I've found that Chinese cabbage grows very well in midwinter because it just doesn't seem to mind those short days, even cloudy ones.

It was all a mystery that first year, because I had no one to learn from who did the same kind of thing I wanted to do in the structure I wanted to do it in. What few books I could find about winter production I read cover to cover, multiple times. They gave me suggestions that got me started, and from there, I observed, learned, and applied the information I accumulated each year. Unraveling a mystery can be fun if you don't let the suspense make you crazy. In my case, that meant too many sleepless nights wondering what could go wrong and how to prevent it. To me, solving the mystery meant eliminating error.

Hanging planter

I sleep better now not because I figured out a flawless system; I made peace with the fact that this mystery never has a final chapter. There is always something new to try and lessons to learn. If that sounds fun to you, too, I hope you go for it. Gardening got me hooked on nurturing plants, but the passive solar winter greenhouse brought out a passion in me to do the impossible: grow great food in a northern winter.

The Garden Goddess greenhouse design ensures a stable growing environment for cold-tolerant vegetables. A planting plan that compensates for the decreased day length of winter solves the challenge of continued production throughout the winter season.

This manual will allow you to learn from my mistakes and triumphs. I won't tell you exactly what to plant and when, but I will tell you what I've learned about the plants I grow so you can apply that knowledge to your own circumstances and start building your own body of knowledge. You can see some basic planting suggestions in the Suggested Planting Schedule section on page 103.

Unraveling the mystery begins with an important realization: winter has three seasons of its own.

The Three Winter Seasons

The three seasons of winter vegetable production are diminishment, solstice, and expansion. For the plants, it's all about whether the days are getting shorter or longer.

Since production in the greenhouse kicks in near the autumnal equinox, we begin with roughly twelve hours of daylight. As you know, each day sees less light until the winter solstice in late December. What some folks don't know is that the rate at which day length changes is greatest during the two weeks before and after equinox. To make sure I explain this phenomenon correctly, I discussed it with my friend Gordon McIntosh, a physics professor and teacher of astronomy at the University of Minnesota, Morris.

As Gordon explains it, the word *solstice* indicates that the sun is in stasis. It is at the peak of its journey north or south across our sky. In both winter and summer, at the time of the solstice, the sun's position in the sky moves very little. The length of the days before and after the solstices vary only by seconds. As we near the point of equinox, when the lengths of day and night are equal, the day length varies by minutes rather than seconds. That adds up over the course of a few weeks. And plants respond to that—both in autumn and in spring.

Each December, when we slip past the winter solstice, you might think that plants will rebound with determination and enthusiasm. Not so. They are actually gaining only a small change per day in early January. They're not impressed. The lengthening of days doesn't really seem to register a noticeable surge in growth until mid-February, when the daily amount of added light increases.

This is not to say that plants stop dead in their tracks during the solstice period. Many winter greens continue their productivity at a modestly reduced rate. But their quality is superior during this time because they like the cool, sunny climate the winter greenhouse offers. The varieties you don't want to grow during solstice are those that are the most phototropically sensitive. Spoken plainly, some foods don't like short days. Some foods don't care. All hail that which fits this "don't care" category. Those are the ones we want to plant in November to get us over the solstice hump.

I have favorites that do well in each mini-season of winter. I think of them as my foundation crops because they grow in enough quantity to form the bulk of that season's greens mix. I plant at least twice as much of the foundation varieties as I do of the others that fill out the mix.

Winter Mini-Season Foundation Crops

Diminishing	arugula and mustards
Solstice	tatsoi and mizuna
Expansion	mixed leaf lettuces

Diminishing Season: Late September to Mid-November

From September to mid-November, each day offers less light. It can be hot in the greenhouse on sunny days, but cloudy ones are more the norm. Arugula is my star in the diminishing time. It puts on weight fast but remains mild because of the greenhouse's cool temps. At this time, my salad mix also includes leaf lettuces, claytonia, vitamin green, red Russian kale, bull's blood beets, chard, and mizuna. As we get into November, I add more Asian mustard greens in the mix, too.

This is a time when humidity is not much of a problem because I can still use the vent system during the day. Also, the greenhouse gets very dry during its summer off-season. Diminishing season is the time for rehydration.

This busy time of winter finds me planting like crazy in the greenhouse every weekend. I add eight new planters of greens every week until the harnesses are filled. It's important to keep up with journaling at this time because it's so easy to forget what was planted and when. You may not care now, but you'll really want to know later.

Monitoring Bugs

The season of diminishing day length is a good time to think about the importance of monitoring what's happening in your greenhouse. I have a fairly dramatic tale of woe to convince you of the need for this particular greenhouse chore.

Regular monitoring is the best way to keep pests and diseases under control because you catch them before they become a problem. Sounds simple, and it actually is, but you have to know what to look for and how to respond when it happens if you're going to prevent a costly and time-consuming crisis. I learned the wisdom in this advice the hard way.

As I prepared for my first winter growing season, I contacted a company that specializes in beneficial insects. If you're not familiar with beneficial insects, I'm about to introduce you to your new best friends. Beneficial insects are among the best tools in your toolbox when it comes to handling greenhouse problems organically. The system is called Integrated Pest Management (IPM), and it's great if you abhor the idea of using toxic chemicals to keep your plants healthy.

Outdoors, IPM includes growing flowering plants near your garden plots that attract the kinds of insects you want in your garden. The good bugs get drawn in by that food source and, in return, help control the pest bugs you don't want. A very nice arrangement.

But nature's system of checks and balances in the great outdoors gets thrown out of whack in the greenhouse since it's an enclosed environment.

So I wanted to hear from the company about how I could take care of the pest problems that were bound to arise.

I was assured that pests are rarely a problem in the first year of operation. "Your first year is your freebie year," the bug guy told me. Of course, what neither of us knew during that conversation was that I would end up planting in the greenhouse that fall, while it was in the final stages of construction. That's just how things were timed, and I didn't think much about it then—I was starting my first winter season!

I'd read about the need to monitor for signs of trouble in various books. I figured my daily stroll to water was a good time to see if anything looked amiss. I didn't take it all that seriously, though. This was my freebie year, after all. Better enjoy it. Well, as you might guess, this sort of thinking made me a prime candidate for becoming a character in what one might call a teachable moment story.

Carol's Aphid Adventure

One fine weekend, I was walking past some young broccoli, and I saw what I thought looked like flakes of ash. Ash? How could ash be in the greenhouse? I leaned in for a better look and realized that what I was looking at were molted insect skins. I felt a panic in my gut. This could only mean . . . I flipped a leaf over, and there they were, my nemeses for the rest of the season: aphids.

How in the world? My freebie year was forfeit thanks to a simple fact of my surroundings. My small, rural town is surrounded by conventional commercial agriculture. One of the favored monocultures in that arrangement is soybeans, and with the beans come problematic quantities of aphids. The aphids are the farmers' problem until harvest. When the soybean fields are harvested, winged aphids go looking for a place to sleep for the winter. Can you imagine their delight when they discovered an unsecured greenhouse full of warmth, moisture, and succulent young green plants? They hit the mother lode! And like all pests, they made use of their good fortune.

I'd read enough about pest problems in greenhouses to know that this invasion was a serious matter. What I didn't know yet was just how effective these girls would be at becoming my worst nightmare. Did you know that aphids are born pregnant? They're ready and able to start adding to your problems the moment they are squeezed out of their mother. Efficient, no? But it gets even better. When the mommy machines start feeling crowded, a few just sprout wings and move to a new location.

So when I spotted my problem, I ordered chemicals. Organic chemicals, yes, but still, the quickie spray fix—I had to get rid of this problem fast. But even if they're organic, sprays are still toxic (they kill living things, after all) and have to be treated with caution and respect. I know organic sprays are a vast improvement over the other ones, but it's still not good to use them in a space as confined as a winter greenhouse. And organic sprays, at least with aphids, have to hit the bug to kill the bug. A winter greenhouse is an intensively planted confined area. Imagine if you will just how much fun it is to be around, under, and behind broccoli plants that require treatment under every leaf. Above them hang nylon harnesses filled with planters of greens. Everybody needs a dousing, and you have to wear a mask, hat, and gloves while you're doing it. Oh, and if you miss just one tiny little insect (which, incidentally, is the exact same color as the plant you're spraying), the whole invasion starts all over again.

The first spray I tried was Neem. It basically slows insects to death. It gets their metabolism to gear back enough so that they basically starve to death because they're too tired to eat. It's kind of creepy, I know, but some of this control stuff isn't pretty.

Well, the Neem spray seemed to knock them back, but after a couple of weeks, the problem was not solved. I knew this because I was applying some of what I was learning the hard way and made daily checks around the greenhouse with a magnifying glass. I had come to know very well what a baby, adult, and winged adult version of an aphid looks like, up close and personal. And they were bouncing back from the spray.

So I tried something else. Soap spray kills aphids because it breaks down the waxy coating on their squishy bodies so they dehydrate.

But again, the soap spray has to give absolutely everybody a shower or the bugs will have the day. I sprayed with soap twice. It did work better than the Neem, but it did not eliminate the bugs. I knew that I couldn't just keep hitting the plants with soap spray over and over again for several reasons. It can burn them at times of full sun, for example.

So I called the company again to ask about bugs. They're more expensive than sprays, but if they could save me from blowing every other Saturday afternoon crawling around my greenhouse with lethal spray, they'd be worth it. The woman I talked to understood that I have a unique environment in my cool winter greenhouse. Outdoors, there are all kinds of insects that find aphids to be quite delectable. Indoors, nobody shows up to the banquet unless I invite them. But the fly considered to be the top killer of aphids, a midge fly, gets snoozy when the days are short and cool, so it wasn't an option. Okay, top killer isn't going to work for me. What else you got?

My advisor suggested that I try some ladybugs. They eat aphids and produce nymphs that devour aphids. Ladybugs are nice to try for a newbie because they are easy to see. Many beneficial insects are so small, they're difficult for us giants to detect unless we're really looking.

And there was another insect she knew would do the trick. A wasp. A teeny tiny parasitic wasp. These wasps detect aphids by smell, and smelling out aphids is all that they're about. They live only a couple of weeks as adults, and reproduction is their sole purpose for being, so when the wasp finds an aphid, she's not thinking, "Supper!" She's thinking, "Nanny!"

The wasp deposits one egg into the abdomen of an unsuspecting aphid. (Aphids are not fast creatures, so it's not a big struggle.) Once the egg is laid, the aphid's fate is sealed. It will gradually slow down and die as the larva in its body consumes it. The wasp pupates in the aphid shell, and when it is ready to leave, it cuts a perfect little circle out of the aphid's abdomen and climbs out, a wasp ready for action. Again, it's gruesome business dealing with pests, but at least this method is natural and completely nontoxic. Also, the wasps do all the work for you.

So I ordered enough ladybugs and wasps for my greenhouse space and waited for them to arrive. A few days before they arrived I did one more round of soap spray. This was recommended to me because the infestation was a bad one. For the wasps and ladybugs to get things under control, there needed to be some food waiting for them but not so much that they would not be able to keep up with it and control it. Also, once the good guys were released, I wouldn't be able to spray anymore because it would affect them, too.

So I did that spraying and put the stuff away, hoping it would be for the last time. I couldn't wait to find out how this whole bug versus bug drama would unfold. My greenhouse was about to become a National Geographic spectacle!

When the insects arrived, I followed the accompanying instructions to the letter. I wanted these girls to feel good about arriving at my aphid party. Some good bugs do better when they're released at a certain time of day. Some come in egg cases that need to be strategically positioned. Some will want to land in a place that's been recently watered so they can get a long overdue drink. It's all in the directions. Before I let the wasps go, I spent some time studying them with my trusty magnifying glass. I wanted to be able to tell them apart from the flying aphids. Didn't want to squish the wrong insect!

After the once-over, I released the wasps and the ladybugs, wishing them much success as they crawled or flew away. Another nice thing about using beneficial insects in a greenhouse is that they're not going to fly off to the next neighborhood, like they can do outdoors. In the middle of winter, nobody's interested in exploring beyond the confines of contained warmth.

The hardest part about using beneficial insects is maintaining trust and patience. This pest control method does not provide immediate gratification. After seeing what aphids can do in a few short days in a greenhouse, it was excruciating for me to let nature take its course.

After a few days, my magnifying glass and I went investigating. It was a mixed review. There were clearly signs of wasp work, but there were also still plenty of aphids. I realized I might not want to squish the ones I did see in case they were already harboring the next round of dedicated aphid hunters. So I took a few deep breaths, backed away, and waited.

I noticed that the ladybugs were not really pitching in. They acted like they'd been sent out on a free Hawaiian vacation—all they did was mate and doze in the sun. In fact, at the very end of the season that spring, while I was ripping out the last of the broccoli, I noticed ladybug larvae on the plants. I hope they had a nice summer outside, but they sure didn't do me much good when I needed them. Ladybugs make nice greenhouse pets because they are kind of cute, they require no care, and they don't really bother anybody. But that's not what I'm looking for in a beneficial insect, so that was the last time I ordered ladybugs.

The wasps, however, proved to be real winners. Because I was able to squelch my panic and let the girls do their job, they did just that. I certainly was glad to be free of the spray duties, but more importantly, the pests were controlled. I never got rid of them completely, but they were kept in check to the point that they did not damage any plants—including the precious new transplants destined for outdoors. Whatever aphids survived in the greenhouse by that point were killed off by the solarizing process of the summer. Nothing to eat and nowhere to hide. Bye-bye girls.

Most often when I tell my lesson-packed aphid story to those who come to hear about our passive solar winter greenhouse, I get the question, "But what happens to the wasps when you're done in the greenhouse? Do they just die?" Well, the adults do, yeah. If they don't hitch a ride on a plant headed out to the gardens or the compost pile, they're toast (literally—in the solarizing greenhouse). But they only live a couple of weeks anyway, so it doesn't make much sense to strive to keep them alive all summer in a dormant greenhouse. It's my guess that most of the wasps hitch a ride outside on plants or find their way out through the vents before the greenhouse is closed up for solarization. I've never seen them flying around after the plants are removed.

The next thing some folks want to know is if I've ever had aphid problems again. Did the Garden Goddess learn her lesson? Well, almost. In my second year, before I brought in the large pots of herbs I'd grown, I trimmed them back, replaced several inches of topsoil, and gave them the once over for any signs of problems. Seeing none, into the greenhouse they went for a winter of harvests. My mistake was in not using my trusty magnifying lens to get a good close look.

I'd certainly learned that I needed to keep a sharp eye out for aphid infestations by that second growing season. I had my magnifying glass out checking for pests as soon as planting began. Sure enough, one day I saw just a couple of aphids on the broccoli plants near the herb pots. I swung my lens over to the rosemary and could see dark red to black little . . . aphids! But wait a minute, aphids are the green of spring grass. What was this new devilry?

Back on the phone with the company, the expert lady informed me that sometimes aphids are affected by the sap they draw from the plants they attack. Apparently, the same aromatic oils that give rosemary its great flavor also change the color of aphids. Tricked again. Ah, but I was already working out the solution.

Another round of wasps was ordered, and this time, I didn't have to spray because I caught the problem early. It wasn't as entertaining without ladybug debauchery, but once again, the wasps did their job beautifully, with no assistance required from me at all. Once in a while, I would see one of these tiny creatures flying from planter to planter, or hanging out under broccoli leaves, but there were never more than a few aphids near them, half of which were usually already dead. Quite often, I saw no signs of either type of insect at all, and that was the true beauty of IPM. The wasps kept the aphids in check and would respond with increased numbers of their own if the aphids gained the advantage. The aphids didn't stand a chance.

By the time my third season kicked off, I was resigned to some sort of aphid infiltration, although I scrutinized the few potted herbs I brought in with what one might call ferocity. I combed the greens. I flipped broccoli leaves (aphids like to hang out on the bottom sides of things for protection) and practically turned the Chinese cabbage inside out searching for the first hint of trouble. I was so focused on ferreting out the first aphids that it wasn't until December that I was willing to entertain the notion that maybe there weren't going to be any aphids that year. Sure enough, thanks to some miracle of happenstance or diligence, I escaped the aphid horde for a season—and then another.

What a treat that's been, let me tell you. Granted, the second year of infestation was far milder than the first, but knowing that you've got plant-sucking cows the size of pin heads crawling all over your food crops is enough to make any grower a little nuts. Now with two years of no infestations under my belt, I suppose I'm getting a little cocky. That sound you hear is the fates rolling up their sleeves to teach the Garden Goddess another lesson in hubris. Good thing I've got the bug guy's phone number close at hand!

Ah, there's one more interesting note to the amazing wasp story. At the beginning of my second season, I was harvesting some of the outdoor crops that were bound for storage down in the cold cellar. One of those crops was winter radishes. They're about the size of a baseball, and they keep nicely all winter. I was tearing off the leaves of the radishes so they could make their way to the compost pile. As I carried an armful of those leaves to the compost, I thought I saw some now familiar things. With closer inspection, I saw that yes, they were the tan, empty shells of parasitized aphids. At some point that summer, aphids had attacked my crop, and wasps had come in to take advantage of the large supply of nannies.

Were these wasps available because they'd emerged from parasitized aphids still clinging to greenhouse broccoli plants, which were now composting near the garden? Or was this simply a drama that had taken place countless times before in my gardens, and I simply was not aware of it?

Now that was an intriguing discovery, and it served as a strong reminder that there are systems and cycles of great complexity in the natural world that I do not always comprehend or appreciate. I find that terribly humbling, oddly comforting, and absolutely enthralling. I will never stop learning so long as I am putting seeds in soil to see what happens.

The pest-free third and fourth years have felt like a big reward for wisdom hard-learned. But I don't kid myself. Each year there will be new lessons that will require every bit of experience I've gleaned thus far. Good thing I wrote it all down in my greenhouse journal!

Solstice Season: Late November to Early January

In late November, the pace is steady and more relaxed. All the outdoor root crops are stored away safely in the cellar. In the greenhouse, planters fill the harnesses, and transplants cover the floor beds. It's too cold during the day to leave the vents open. They have to be secured against the fierce cold winds of winter storms. On cloudy days, the temperature in the greenhouse rarely goes above 60 degrees F (16 degrees C).

Cool and cloudy days become less frequent in solstice. It's colder outside but sunnier, too, and that makes it lovely in the greenhouse. Even on days when the outdoor temp is in the negative teens, with windchills in the minus 20s F (minus 30s C) or worse, it gets hot enough in the greenhouse that we open the door into the packing shed to help cool things down.

Incredible, isn't it? But it's the truth. In fact, today the high outdoors was minus 13 degrees F (minus 23 degrees C). Winds in the morning offered a windchill of minus 28 (minus 33 degrees C). But it was over 90 degrees F (32 degrees C) in the greenhouse by noon. Even I didn't expect it to be that hot. The trick is to get things cooled down without overstressing plants with a rapid temp change. But cooling is required, because these are the same plants that are most happy in temps around 60 to 65 degrees F (16 to 18 degrees C).

The main problem to watch for is the effects of too much humidity. It's some-times impossible to vent directly outdoors because of the extreme weather. I have to keep my eye on my watering practices to make sure that only the plants that need moisture get it.

Many plants simply do not grow during this time of short days: for example, spinach, bunching onions, kohlrabi, and salad radishes. If these vegetables are not already in the ground and near harvest size, don't bother planting them until February. They'll just sit there and wait—which is not exactly a healthy thing for a plant to do in a cool, high-humidity environ-ment. On the other hand, Chinese cabbage does remarkably well in solstice season. Pac choi does okay. Both taste terrific in the solstice season. Unlike outdoors, these plants are not subjected to the attentions of insect pests, nor do they get tough and bitter from heat. In fact, the Chinese cabbage is so sweet at solstice, I use the juicy leaf ribs as an excellent substitute for celery. They are never bitter or stringy.

In the planters, mustards rule at solstice. Mizuna is the star, but red giant also does well and provides lovely color in a salad mix. Garden cress is another good solstice green, and its peppery zip is a welcome addition to the mix. Tatsoi is a great solstice season grower, too. It's in the brassica family and makes small spoon-shaped leaves that are a deep green and taste some-thing like spinach. Tokyo bekana is the most recent addition to the solstice planting plan because it just grows like crazy.

Best Vegetables in Solstice

Chinese cabbage	Garden cress
Pac choi (bok choy)	Tatsoi
Mustard greens	Tokyo bekana

Less Is More

The time of solstice, when days are short and every minute of sunshine is precious, is a good time to consider the basics. For example, lots of folks ask me why we don't use artificial lights to increase the productivity of our green-house in the depths of winter. Well, sure, I thought about it. Commercial growers use them to raise tomatoes and cucumbers in winter. But I agreed with what I read in Eliot Coleman's book *The Four Season Harvest*. The more I rely on the sun and less on fossil fuels, the better my business will be able to handle the challenging times that face us as the stores of those finite fuels are drastically reduced. The price of fuel is only going to go up over the long haul, while the sun's energy is still free for the gathering.

Lettuces

The lights required to help vegetables grow in the dark of winter are expensive to buy, and they suck up *a lot* of electricity. That means I'd have to charge more for my produce. Would the resulting increase in productivity make up for the cost I would have to charge back to my shareholders? Maybe, maybe not. But I do know that I like being able to say that we use very little fossil fuel right now to grow the crops we have.

That lighting situation may change. I have heard some rumblings about LED lights for the greenhouse that could be powered by solar batteries. That's cool technology, and I do love hearing about such developments, but until it all gets sorted out, I think I'll stick with the less-is-more plan.

Plenty of tasty veggies grow just fine in the winter without the added expense and hassle of artificial lights. Why not learn more about those crops and how to grow them, instead of forcing some other plants to perform in their off-season? I've tasted winter greenhouse tomatoes, and I'm sorry, but they taste nothing like the ones I plucked off my vines in August. So, in the winter, I make do with frozen, dried, and canned garden tomatoes, and I feast on the finest greens in the land because that's what the winter greenhouse—with miniscule props from dino-fuel—does best.

That's not to say I'm unwilling to test my ideas about how to improve this winter growing business. I was awarded a research grant that allowed me to test soil temperature and planting guidelines, as well as new greens varieties. You can read about the results on pages 93–94. Some of the results will probably surprise you.

Expansion Season: Mid-January to Late March

The expansion season is the light at the end of the tunnel. At some point in early- to mid-February, I feel a shift in the greenhouse. It's silly to give plants credit for an attitude adjustment, but within a week's time, I notice greater vigor in plant growth. What a relief!

This shift prompts me to make some adjustments in the greens I choose to seed in the planters. I sow more mixed lettuces because they handle the increasing temperatures best. Their various shapes and colors keep the mix of greens interesting. They are sweeter than the greens we get used to in early winter and midwinter, but that seems appropriate as hopes for spring begin to rise in our expectant hearts.

Mustards are inclined to bolt when longer, warmer days bump up the temps in the greenhouse. Their flowering stalks are actually edible, as are the yellow flowers that smell like daffodils. But when this happens, the plants are more than done with production and need to be replaced. At this time of year, I may only get a couple of cuttings from mustards before they bolt.

Arugula is also good for only two cuttings before it needs to be replaced. Tatsoi will give three if it doesn't get too much heat stress. Both of them just don't like this greenhouse season very much. I place such varieties that are less heat tolerant in the lowest hangers of the planter slings. The lettuces are joined by baby kale and collard, beet, and chard greens because they handle warmer weather better than the other greens do.

Fresh outdoor air provides a welcome relief from winter's lockdown. It soon becomes a necessity, for as temperatures rise outdoors, the greenhouse must be vented to stay cool enough to keep the crops happy. It's a good time to work some salad radish seeds into any blank places in the floor beds for a late spring harvest.

Expansion is an exciting season because transplants are being sown for the outdoor gardens. Leeks and celeriac benefit from an early start in the greenhouse. Onions also are seeded into soil blocks. Finally, in late March, I give the tomatoes and peppers an early start. I'll transplant them later into larger pots, which will give their roots plenty of room while they wait for outdoor soils to heat up properly.

The end of this season is all about transitioning to outdoor garden plots from the confines of the greenhouse. Final harvests leave open space in the floor beds for the expanding number of transplant flats.

One of the many great things about having a greenhouse is that you'll be able to grow wonderful transplants for yourself. During the expansion season, as the plants that feed our share members cycle out of the growing space, I have room to start seeds of my favorite flowers, herbs, and veggies for my own summer garden.

It's fun to try out varieties of flowers or vegetables that aren't commonly available at nurseries. I've made a few bucks on the side in springtime selling planters of flowers to my friends. I find nice planters at rummage sales for next to nothing and stash them away for spring planting the next year. I have also sold special varieties of peppers, tomatoes, and herbs. This ends up covering the costs of seeds for myself along with the occasional indulgence in perennial flowers I can't resist.

Spring is an exciting time of year. For most folks here in the north, that pleasure doesn't happen until April or May. In the greenhouse, it comes to you in mid-February, and that is a welcomed burst of life indeed!

Mid-April is the last delivery of the winter CSA. At that time, I am usually very happy to turn my attention away from the confines of the greenhouse to the much larger growing beds outdoors.

Spring:
Phase Down and Gear Up

Spring is the beginning of the end of the growing season in the greenhouse. It's the time when I give the greenhouse's packing shed a good cleaning and reorganizing to get ready for the transition to outdoor gardening needs.

The Packing Shed

A packing shed is what it says it is: a basic space where you deal with the harvest and prepare it to meet the public. This is not a fancy place, it is a practical place. Your packing shed's layout will be dictated by what you are growing and for whom. For an operation like mine, collapsible waxed boxes are the standard container for transporting a mixture of garden produce. I've got to have a space to store them, along with produce bags, scales, rubber bands, work gloves, face masks (for when I have a cold or am working with fine-particled soil mixes), and various tools. I also need counter space that will allow for basic processing and that can be washed down for cleanliness.

I'd like to brag that I did all sorts of research on designing the most efficient layout for this space—but I didn't. I had a general idea of how work would flow and designed the room layout to make that happen. I did save myself some serious money by hauling home used cabinets from friends who were remodeling their kitchen. What a find those were!

I think the layout of the packing shed becomes much more important if you are going to be sharing the space with others. I most often work alone when I am packing shares, so I do not have to dance around the work of others. If you will have an operation larger than mine, I suggest visiting as many summer CSA farms as you can to see how they process their produce. You'll appreciate how efficient farmers and their workers can be, and the clever systems they create. Work for a day with the packing crew, and you will get a feel for the work flow. You'll come away with lots of ideas on what you want in your own work space. Be sure to visit CSAs or market farms that are similar in size to your own operation. Solutions are quite different for small, medium, and large vegetable farms. Find out what they can teach you.

hardening off

the gradual process of acclimating a greenhouse transplant to the outdoors before it is planted

With the end of the winter season, I store all the empty planters and put transplants outdoors to harden off. The packing shed becomes more of a tool shed, where I keep the seeds, supplies, and tools I need for the garden plots.

Toward the end of spring, our greenhouse is shut down and largely forgotten during the busy outdoor growing season. I have tried in the past to grow heat-loving crops in the greenhouse, but it hasn't worked. It just gets too hot in there—even for melons and hot peppers. All the design elements that keep heat in the structure make it hard to cool it off after May comes around. Blossoms fall off the plants, and they just swelter. But when we get a chance to build another greenhouse, I think I'd like to risk some added ventilation and have another go at the melons.

The Rest Is Up to You

Now you know what happens in the Garden Goddess greenhouse every winter. You can take the knowledge we've gathered and apply it to your own version of winter production. It's not a matter of whether it can be done—we've proven that. The real question is how you'll take our information and adapt it to your own circumstances. You may be in a city or on a farm. You may have a local grocer who wants to buy wholesale from you, or maybe there's a fancy restaurant in town that'll buy every ounce of greens you grow. There might even be a school near you that would love to offer your winter produce as part of its lunch program. Perhaps you'll choose to go the CSA route because of the customer interaction it provides.

How you market the produce is up to you. How you modify our design and production methods is your own adventure. You've got our basic production methods and experience in observing how plants respond to the day length changes of winter. These tools in your imagination's toolbox will help you fashion the greenhouse that matches your own goals and dreams. You *can* do it.

Of course, it helps to have some physical tools as well—especially when you're growing vegetables. Below I've listed some of the essentials you'll need for greenhouse production.

Your Greenhouse Tools

The tools you use in the greenhouse should not be used outdoors. Having tools designated only for the greenhouse helps prevent transmission of disease or pests from the outdoors. And when you get into the habit of having greenhouse-only tools, you'll never find yourself in the middle of a project in the greenhouse wondering where the shovel went.

ITEMS	NOTES

Tools for Working with Soil

- Wheelbarrow. *Or some other portable large container for mixing soil.*
- Shovel. *To help incorporate soil amendments into raised beds.*
- Hoe. *To help in combining soil mix ingredients, smoothing raised bed surfaces, and weeding.*
- Rake. *For surface-mixing of soil amendments and also shaping raised beds.*
- Soil thermometer. *A good one. You'll want to keep track of the temperature in the raised beds.*
- Soil block maker. *There's only one thing this is good for, but nothing else will do.*
- Plastic flats. *To hold soil blocks—at least a dozen, preferably more.*

For Working with Plants

- Planting information stakes. *I make them out of empty plastic yogurt and other containers.*
- Hand tools for working with transplants in the floor beds. *Use whatever you like and what feels good in your hand. I prefer a fork over a trowel. I have a long-handled fork that's great for extending my reach, and I have a smaller double-sided tool that*

ITEMS   NOTES

is a fork on one side and a trowel on the other. It's
good for making holes for transplants.

- Scissors. *For thinning seedlings.*
- Watering can. *It comes in handy for spot watering
  and pouring water into flats of soil blocks. The best
  kind is a lightweight plastic one with a fine spray rose
  that can be taken off.*
- Watering wand with adequate hose. *For getting to
  all corners of the greenhouse.*

For Harvesting

- Several sharp harvest knives and a whet stone to
  keep them that way. *Don't use your harvest knives
  for anything else. I keep other knives in my packing
  shed for trimming storage crops. That heavier kind
  of work dulls a harvest knife quickly, and you don't
  want that.*
- Lots of different kinds of harvest containers.
  *Again, plastic works best in the humidity of the
  greenhouse. It's light and easy to clean in bleach
  water. I find useful harvest containers at dollar stores.
  They can have an open weave to them like a small
  laundry basket or a sieve design so long as the weave
  is not so large that greens' leaves can fall through.*
- Salad spinner. *You need a commercial one because
  a kitchen spinner can't hold up to the beating it will
  take doing the volume of greens you grow and harvest
  every week. They're not cheap, but if you wind up
  with the kind of problems I had in the first year (dead
  aphids that had to be washed off baby greens), it's not
  optional. It's required. Nobody wants to get a bag full
  of dirty greens. Greens sell the whole winter CSA*

ITEMS   **NOTES**

*for me. They have to look as good as they can. They
are small and fragile and can't take a lot of rough
handling. A good salad spinner removes excess
moisture after cleaning but doesn't beat up the greens
doing it. Check out the one offered in the Johnny's
Select Seeds catalog. It's worked well for me and other
CSA farmers in my area.*

LEARNING FROM EXPERIENCE

What to Grow and When

The raised beds of the winter greenhouse allow us to grow cold-hardy vegetables that require more space for roots than the planters provide. I am still experimenting in the raised beds to discover what crops grow best there. I will share with you what I've learned so far in the section below on raised beds. Next, we will explore the fascinating and, dare I say, exciting world of greens grown in planters. Below I discuss some of the best varieties I count on to keep me and my shareholders enjoying a tasty array of fresh salad greens all winter.

Raised Beds: Large Crops

Broccoli

Pros: Produces a premium product.

Cons: Takes up a lot of space for the amount of harvest produced. Aphid magnet.

Broccoli harvested from the winter greenhouse is wonderfully tender and sweet. It lacks the strong flavor or bitterness that summer broccoli sometimes has. Greenhouse broccoli has never been hassled by a white butterfly or its accursed green worms (that cause us all to look carefully before we bite into our garden-grown summer broccoli heads). The most succulent, flavorful broccoli I've ever tasted has come from my greenhouse, and that's not just because *I* grew it. Greenhouse, take a bow.

The best broccoli variety I've tried in winter is DeCicco, an Italian broccoli that puts out a small main head and then many side shoots. Once a plant has started putting out broccoli shoots, you'll be harvesting them steadily until spring. If you prefer a variety with a larger head, I found Arcadia to be good in winter. It does produce some side shoots, just not as vigorously as DeCicco.

The challenge I'm still working on is how to get broccoli to produce sooner in winter. So far, my DeCicco and Arcadia haven't been ready to harvest until late January. Part of the reason is that broccoli can't be transplanted into the greenhouse's floor beds until mid-September, when the heat of summer subsides. This means seeds get planted in soil blocks in mid-August. I keep them outdoors in the shade, where it's cooler, until they sprout. Then I get them into more sun, protected from pests with row cover. It's a bit of a fuss and runs the risk of bringing pests into the greenhouse, but it's my only way to have broccoli ready for September planting.

Broccoli takes up a lot of space as it matures, but I grow it because share members want it so much. With a good side shoot producer, I get at least a January to April harvest from it. I interplant other crops with the young broccoli in autumn to make the most use of the raised bed where it grows. Pac choi is a good choice for this duty. By February, broccoli shades out any other plants in the bed, so plan accordingly.

I've also found it helps maintain the health of winter broccoli to snap off the bottom leaves once the plant is large enough to start setting heads. This slow grower is better off losing those bottom leaves that touch or lie on the soil surface. By the end of a long winter, such foliage is prone to fungus or mold problems. Snapping them off as the plant grows is also a great chance to inspect for pests. Take a hard look at the underside of pruned leaves. If these leaves were stressed, you can bet that's where pests will go.

Broccoli Raab

Pros: Vigorous grower. Can eat the whole plant.

Cons: Also an aphid magnet

This is the crop I watch most diligently for signs of aphids. Like regular broccoli, it seems to be a favorite of the aphids that find my greenhouse.

I am finding that broccoli raab does even better than its cousin in winter production. If you've never had raab, it's in the broccoli family, producing stalks with flower buds something similar to but smaller than typical broccoli heads. Raab is grown for its leaves and stems as well as its florets. The whole works can be steamed and added to soups, stir fries, or (my favorite) pasta with olive oil and garlic. Broccoli raab has grown very well for me in the diminishment time of the greenhouse; its deep green leaves seem to sing of healthful minerals and vitamins. I haven't settled on a variety yet, but I have found that I can get two crops of raab out of the same growing bed in one season. With good planning, I can rotate raab with other crops and come out with three harvests of it in a full winter season. That's a lot more edible produce than broccoli provides.

Get the broccoli raab started in mid- to late August so it can be transplanted in the greenhouse as September days and nights begin to cool. You really don't want these plants in a greenhouse that can't be cooled to 80 degrees F (27 degrees C) during the day, so keep them outdoors and protected with row cover until then. If you're worried about insect pests, give the transplants a soap-spray treatment before bringing them into the greenhouse. Do I even need to say that you must follow the directions of any insecticide and use it with respect and caution?

Chinese Cabbage

Pros: Fast grower. Tolerant of "cold spots" in the
          greenhouse.

Cons: Can't think of a one.

Chinese cabbage is not the star of the winter raised beds, but it is a workhorse. The whole plant is harvested, so there's no waste. It can be planted thickly and then thinned for successive harvests. The older heads are just as succulent as the young ones.

I like the Chinese cabbage variety named Minuet. It's an open leafed cabbage with an attractive green color that has a mature look to it even when it's small. I start it in flats of soil blocks in early September. More get planted in mid- to late December to replace the first transplants when they're harvested.

Pac Choi

Pros: Grows well all winter. Whole plant harvested. Good for keeping all greenhouse spaces productive.

Cons: There's nothing about this crop I don't like except that it doesn't patrol the growing beds to slay aphids while I sleep at night. Oh well, can't have everything.

A rose by any other name: pac choi, bok choy . . . same plants, same great taste. The crunchy, juicy stems are my favorite veggie to enjoy with dips, and the succulent green leaves? Well, I chomp them raw. They're also wonderful in stir fry, of course, especially one that involves spicy peanut sauce. The pac choi variety I favor is Mei Quin, a small variety with deep green color in the leaves and creamy white stalks. I grow them the same way and at the same time I do the Chinese cabbage.

Chard

Pros: Loves the winter greenhouse. Colorful stems and deep green leaves are welcome additions to the winter harvest. Good producer all season.

Cons: Have to watch for dampness problems developing on soil surface beneath chard's big broad leaves.

In my fourth year (2008–2009), I decided to try chard in my raised beds. Along with kale, the chard went into the beds in early October, and I harvested the first cutting in late December. This crop was a great success, and I will continue to plant it. After the first cutting, the plants just kept producing like crazy for the rest of the season.

The leaves are more succulent and tender compared to the chard I harvest from my fall gardens outdoors. I eat the winter chard leaves raw, usually putting them to service wrapping up whatever rice and bean, meat and tomato, or scrambled egg concoction I've mixed up. The wraps can be warmed or baked, too. Talk about a versatile and nutritious food!

Bright lights is an especially attractive variety because of all the different colors. The trick is to keep harvesting from the bottom. Also, don't cut the center's growing tip. I let each plant keep a few young leaves when harvesting the rest. Cutting most of the leaves gives the soil beneath the plants a chance to air out and provides an opportunity to clear out any moss or small mushrooms that might get a toehold under those elephant ear–shaped leaves.

Kale

Pros: It tolerates cold spots in the greenhouse.

Cons: Insufferably slow grower.

I tried dinosaur kale in 2008–2009, and I don't think I'll plant it again. It was a very slow grower. I managed only one harvest out of one planting, as compared to three cuttings of the chard. I'll most likely try some other kale varieties to see how they do. I grow red Russian kale in the planters, and it makes a nice baby green in the salad mix, but I've never much cared for it in its larger form. Fortunately, there are a few other options available in the kale world; I'll test those before I conclude that it's not worth the bother.

Hakurei Turnip

Pros: Easy to grow. Tops make a great braising green so nothing is wasted. Really tasty, quick-growing root crop for the greenhouse.

Cons: Doesn't grow much during solstice season. Needs to be planted in early diminishment season or mid-expansion season only.

Hakurei is a salad turnip from Japan that is all white and grows to a size somewhere between a ping pong ball and a racquet ball. I also grow it outdoors for a fall crop. The leaves are edible, which means nothing goes to waste. The root's sweet, crunchy flesh is great raw, diced, or shredded on those lovely winter greens.

I broadcast hakurei seed in a raised bed, cover them with soil mix, tamp it down with a flat metal rake, and water. When the seedlings emerge and grow their first set of true leaves, I thin them to be about a couple of inches apart. I can usually get two harvests out of this seeding by taking the biggest ones first, leaving the rest for a couple more weeks to put on some weight. When they are harvested, I leave the tops on. If they won't be used immediately, the greens should be cut off and bunched separately, but I like being able to send them off whole to the shareholders as a little reminder that we are rarely blessed to have such fresh local produce available to us in the blustery midst of a northern winter.

Radishes

Pros: Crispy and not too fierce in flavor when grown
in the greenhouse.

Cons: Greens are not all that great. Only seems to produce
reasonably quickly in later part of expansion season.

I've had my best luck with icicle radishes in the winter, but none of them are excited about growing until after the spring equinox, so it's a race to seed them and get a harvest before the greenhouse delivery season is ended (mid-April). Still, they can be a tasty different veggie to add to the last shares, and they're a good stir fry addition. I seed them the same way I do the hakurei turnips, but I don't bother with the thinning after they sprout. They get pulled when they're full-sized, granting more room to the ones around them not ready for harvest.

Unsuccessful Attempts

You should always persevere to test new crops for success, but be prepared for some failures. Crops I have tried to grow that failed me miserably include bunching onions, kohlrabi, spinach, and mache. This surprised me because these crops don't mind cool temperatures at all. They are strongly recommended for season-extension hoop-house production. In that system, the crops are planted in late summer in hoop houses that later protect them from autumn's biting winds and chill. The mature crops sit with their feet in cold soil, not growing but not freezing, until harvest. It's a pretty clever way to get the effect of a refrigerator without the expense of electricity.

But alas, these crops seem to have a phototropic sensitivity that shuts them down for the winter. In my greenhouse, where I want them to be growing plants, not dormant ones, they shut down. Spinach that I plant in October can't just sit in a raised bed until late March, when it feels like growing again.

It's a waste of precious growing space to let plants snooze in the winter greenhouse. I need them to produce! Fortunately, many can. But spinach is a sleeper, I'm sad to say. And so are the rest I listed. One year, I wasted the growing space of an entire raised bed trying to get kohlrabi to grow. It not only refused to swell into edible bulbs but lethargically suffered through the winter only to suddenly bolt to flower stalks with the first hot temps of spring. No more kohlrabi in my greenhouse!

Go ahead and try them if you want—maybe you'll have better luck than I did. But after a couple of disappointing attempts, I tend to move on with my search for the best producers in my particular environment. Let me know if you figure out a way to make them grow!

I should in all honesty admit that I'm not above ignoring my own advice. I had rotten luck in my first season when I tried to grow head lettuce. It wouldn't produce for me, and so I dropped that from my list of crops for the raised beds, figuring all the greens in the planters more than made up for a lack of head lettuce. But I've decided that it's time to revisit that crop. Next year, I'll experiment with a couple of varieties to see if I can make a go of it. The quest continues!

Trying Out Warmed Soil

I did an experiment during the 2007–2008 winter growing season on the less-is-more theory of winter production. A USDA Sustainable Agriculture Research and Education (SARE) grant gave me the opportunity to research which greens varieties do best in solstice season (see page 83). As another component of that grant, Chuck built us a raised bed that can be heated. He placed a metal cloth with heat tape attached to it six inches below the surface of the soil. This allowed us to keep that soil—at root depth—at 65–70 degrees F (18–21 degrees C). The rest of the beds are more like 50–55 degrees F (10–13 degrees C) in midwinter.

Going into the experiment, I thought that we would see some increased productivity in the heated bed and that we'd have to determine if that added growth was worth the expense of heating the soil. The plants had other ideas. I grew broccoli, pac choi, and Chinese cabbage in the heated bed and in the regular raised beds—all planted and harvested at the same time. I measured the harvest height of the plants and the weekly harvest weight. Imagine my surprise when the plants in the regular beds consistently outperformed the plants growing in warmed soil. One week, the Chinese cabbage in the regular beds weighed twice as much as the Chinese cabbage in warmer soil!

The message is clear: plants that do well in a winter greenhouse do not appreciate being coddled. They like their feet cool. Providing a system they don't need would be a waste of effort and energy. Sure, it might work for other crops that do like to have warm feet, but they generally also like long sunny days and warm air temps—something the greenhouse is shy of during December and January.

Less is more. You can grow abundant harvests in a low-tech system that's easy on your busy schedule and on the environment. Why fight such a good thing when you've got it?

Planters: Great Gathering of Greens

And now for the stars of our show: the mighty micro greens!

The fresh greens that come out of a winter greenhouse are a valuable, nutritious, and appreciated food. More than once I have had shareholders admit to me that they're in it for the greens. They enjoy the other vegetables found in their weekly shares, but they anticipate and savor the greens above all else. There's nothing like them. The stuff in cellophane bags from California or beyond just doesn't come close.

It behooves you, no matter how you market your greenhouse harvest, to maximize the amount of greens you produce in your structure. We find the planter system we've used to be a great way to raise and harvest greens grown to a height of 4 to 8 inches. And after that, it's all about knowing which varieties to grow during the various stages of winter.

In the planters, we grow over two dozen varieties of greens. As I test them, I note the growth habits, strengths, and weaknesses of each variety. Here are my observations after growing them for four seasons.

Arugula

Pros: One of the best growers all winter, quick to germinate, great taste

Cons: Older plants get woody stems; quick to bolt in warm temps

Ah, the queen of the winter greenhouse: arugula. I love, love, love this green. It pops out of the soil typically in just three days and usually within three weeks is ready for a first harvest. That harvest is always the thickest and best-looking, but I know there will be three before it's done. And I just keep planting more.

Arugula doesn't mind midwinter one bit. What it does mind is heat. This green will bolt in a heartbeat, especially in spring, when the temperatures get toasty. When that happens, arugula becomes bitter and tough. If the stems are beginning to look a little woody, you probably have a planter that needs to be dumped and replanted.

Arugula can take thick seeding just fine, and it gets me through that tricky solstice season every time. It's also a widespread favorite among share members. Everybody seems to like the taste and texture of this green. It's certainly milder than what you're used to out in the spring or summer garden. Arugula has proven itself as a real workhorse in the dead of winter.

Bull's Blood Beets

Pros: Wonderful color, unique flavor, good quality in all harvests
Cons: Slow grower

This one's a real slow grower, all winter long. It makes up for this by producing the most beautiful and flavorful greens I offer. The colder it gets, the deeper the rich burgundy color of these baby beet greens. They look like rubies in the salad mix, and a little goes a long way to give that colorful

Chard

effect. Like claytonia, I space out my bull's blood beet plantings because they do slow down the harvest schedule with their pokey growth habit. But they don't bolt, and they maintain a steady flavor quality right through a fourth harvest. They're definitely worth keeping in the mix.

In 2008–2009, I tried a variety of yellow beets in the planters to see if it would work out as a baby green. I've found that the yellow beets I've grown outdoors have yielded the best quality cooking greens out of all the beets, so I hoped as a baby green it would also shine. Unfortunately, it acted like spinach (which it is related to, after all). It stopped growing when it was a couple of inches tall and just stayed that way through the winter. I'm going to stick with the bull's blood variety from now on.

Bright Lights Chard

Pros: Wonderful color, taste, leaf size, and staying power

Cons: Slow grower as a baby green

I didn't experiment with chard as a baby green until my second year of greenhouse production. I was so busy growing it outside in the autumn gardens that I figured everyone had their fill of it in the fall. But I am completely sold on bright lights chard in the greenhouse planters now.

First off, there are all those great stem colors—yellow, cream, pink, red, orange—that carry into the veins, too. Add to that a pleasant mild flavor and harvests that just keep coming and coming. I've had five cuttings off some plantings of chard, although four is the norm.

They are slow growers, however, at least when they're getting started. After the first harvest, they speed up a little. They really are trouble-free in the greenhouse, and I find myself seeding planters of chard more often with each passing season.

Claytonia

Pros: Appealing leaf shape, texture, and taste

Cons: One of the slower-growing varieties

Claytonia's succulent heart-shaped leaves have a fresh, sweet taste I love. Waiting such a long time for that first cutting does drive me crazy, but after that, it grows faster and will even give four cuttings without too much fuss. It just takes its time germinating and getting started—no matter what season I try.

To compensate for the fact that claytonia will be nowhere near harvest when all the other planters I seed at the same time are ready for the knife, I plant only one planter of claytonia at a time, and I spread the plantings at least a couple of weeks apart in the schedule. That way, faster growers can help fill in the gap while claytonia takes its time getting to salad size.

Garden Cress

Pros: Grows well in all seasons; terrific zippy flavor

Cons: Older leaves yellow and need to be culled after
harvest

A little cress goes a long way in the salad mix, but I wouldn't want to go without it. Though not as productive as greens, it's a tasty seasoning for the mix all through the winter.

Cress is also a good place to go looking for the first signs of an aphid infestation. Aphids seem to love this plant and are easy to spot on its long stems.

Lettuce Mixes

Pros: Quick, even growth (varieties reach harvest size at
the same time); can handle heat better than mustards
and brassica varieties

Cons: Not much to them; some midwinter slowdown;
can develop rot problems if seeded too thickly

I've tried a bunch of lettuce mixes from a bunch of seed companies, but my favorites are wildfire and allstar gourmet, both from Johnny's Select Seeds. The red lettuce varieties in the wildfire mix get richer in color as the weather cools. I usually get three good cuttings from these mixes throughout the winter. After three harvests, especially in the expansion season, the baby lettuces can develop a bitter taste that is not pleasant. Leaf shape and color vary in these lettuce mixes.

Be cautious not to overseed planters of mixed lettuce. If they are too crowded in the humid cool weather of the greenhouse, they develop a problem I refer to as "meltdown"—they just plain rot. Not allowing baby lettuce to get overgrown before cutting also helps prevent meltdown. At harvest, make sure you cut them down as close as possible to the soil level. This gives things a chance to air out a little before the next round of leaves shoots up.

Baby lettuce doesn't add a lot of lift to a greens mix, but its sweetness is a nice foil to some of the fierce-flavored varieties. Also, in early spring, when things are heating up in the greenhouse, lettuce plantings take up the space at the top of the planter harnesses, where it gets warmest. They don't bolt as quickly as other varieties—they can take some heat and keep their quality.

Red Russian Kale

Pros: Beautiful, deeply serrated leaves; unique color and flavor; handles cold very well; adds heft to the mix

Cons: Germinates and grows more slowly than other varieties

I think this is one of the most attractive members of my winter greens mix. The frilly silvery-maroon leaves have reddish-purple veins. They are firm, substantial leaves, so they provide weight and lift to the mix as well. And they taste great.

I grow this variety throughout the winter season even though it germinates and starts growing at a slower rate than most of the others. I find that the growth rate picks up during the last part of the winter season, and the plants handle the heat of spring. I've never had red Russian kale bolt on me, but too much hot weather does affect flavor quality, so eat a few leaves before harvesting the last of the three cuttings you'll get from this variety to make sure the taste is still good.

Komatsuna

Pros: Similar to tatsoi but leaves are larger and longer, with less curl

Cons: Doesn't grow quite as fast as some of the others

Komatsuna is one of those dark green leafy cole crops that like the cool, humid greenhouse very much, thank you. When I started out, my main greens variety of this type was tatsoi, but I soon found that komatsuna has all the advantages of tatsoi like color, flavor (much like spinach in taste and texture), and cold-hardiness, and also produces a slightly oval leaf shape as opposed to the tatsoi's circular shape.

Even if none of my share members can tell the difference between tatsoi and komatsuna, I can. The leaves are often two times larger than tatsoi's in the same growing time. The plants aren't the fastest growers, but in cool temps they'll easily give four harvests before they're done. Watch them in the spring, though. After the second cut, they may bolt if they get hot.

I tried a red version of komatsuna in 2008–2009. It is a striking purplish color, but it grows even more slowly than the green variety. I do like adding color to the greens mix, and this one kept its quality through four cuttings, so I'll grow this one again next year, but I'll start it early in the season to see if that speeds up its first harvest date.

Minutina

Pros: Crisp, fresh taste, unusual shape, steady grower
all season
Cons: Rots if planted too thickly; can be mistaken for grass

Minutina is great because it fools so many people with its green, grassy leaves. They wonder at first inspection if I've lost my mind and tossed grass into the salad mix—it really does look like a mat of grass before cutting. But minutina leaves are succulent and offer a sweetness to those who dare to try them.

Minutina doesn't win any races to the harvest finish line, but it's not at the back of the pack either. It shares a characteristic with lettuce: if it's seeded too thickly in planters, it's prone to meltdown (rot). Sometimes in the solstice season, thickly planted leaves just don't get a chance to dry out. After three harvests, this one's done: the leaves start to toughen and lose their sweetness.

Mizuna

Pros: Great leaf shape, color, and flavor; fast germinator
and grower
Cons: Will bolt quickly if greenhouse gets too warm

Mizuna is a real champ. It germinates in only a few days, even in the coolest times solstice season can offer. It also has deeply frilly leaves and is one of the milder members of the mustard family. It's got a pleasant medium-green color, and it helps give some loft to the greens mix.

Of course, no greens variety is perfect, and mizuna has one personality flaw. As soon as it feels the least bit warm, it bolts. I've seen it happen after just one day when I didn't get the warming greenhouse cooled fast enough. When it begins to bolt, mizuna is done.

I forgive mizuna this quirk because it has helped me make it through cloudy midwinter weeks when nothing else would push up through the soil. I know I can depend on mizuna when times are tough, so I cut it some slack in hot weather. It's a fair trade, and I know if I monitor things correctly, I can coax it into reasonable production from the beginning to the end of the winter growing season.

I've also found that if I let a planter go and it flowers, its small yellow flowers smell daintily of daffodils, so I cut the stalks and put them in a vase to enjoy in the spring greenhouse. The flowers are also edible and can be added to spring salads for some visual appeal.

Osaka Purple Mustard

Pros: Great leaf color and size, good flavor, good growth

Cons: Flavor can get very peppery late in the season

This mustard is a terrific green because it has some hot zip to it (but not overly so) and purple veins that look pretty in the mix. It grows pretty well, even in solstice season, and makes good-sized leaves, too. You do have to taste-test it in spring, because higher temps in the greenhouse can mean higher heat in the mustard greens. Some folks really like that, but I prefer their milder winter flavor.

Red Giant Mustard

Pros: Nice leaf color and size, mellow mustard flavor

Cons: Medium-slow grower

Red giant is another of the mustard greens that provides great color and texture. It produces larger leaves faster than most other varieties, but the bigger they get, the stronger their flavor, so watch them if the fierceness matters to you. Again, I have share members who find this to be a winning feature of a winter green, so I keep this one around until spring, when it just gets too strong.

Tatsoi

Pros: Nice deep green color, succulent texture, great mild cabbagey flavor

Cons: Medium-slow grower; prone to bolting in spring

Tatsoi quickly became my replacement when I realized I couldn't grow spinach worth a darn in the winter. Its leaves tend to get a sort of scoop shape to them, and the white stems are crunchy and juicy. The only down side of this steady, slow grower is that it doesn't care much for heat. In spring, after just two cuttings, it may well decide it would rather make flowers.

Vitamin Green

Pros: Larger leaf size with dark green color

Cons: A little slow growing after first harvest

Vitamin green may well be my favorite of the cole family of baby greens. It makes an oval leaf similar to komatsuna's but very dark green. It seems to grow a little faster than tatsoi, although it is certainly not in the fast category of growers. The taste is terrific, and it seems a little slower to bolt than tatsoi.

New Kids on the Block

As part of the USDA Sustainable Agriculture Research and Education grant I was awarded for 2007–2008, I tried many new greens varieties. The purpose of the grant was to work on ways to increase production in the winter greenhouse. One important area I continue to work on improving is harvest during the solstice season. Even the best of the greens slow down when days are shortest. The thing to discover is which ones actually produce during solstice season so planting schedules can be modified accordingly. I don't want to be filling planters with the slowest stuff I grow right before the season of shortest day lengths hits.

The results of that research were useful, and I'm excited to apply what I've learned in the coming seasons. While I can't profess to know these new varieties as well as the others, I do want to share what I've discovered thus far. Most likely, I'll have more to say about them down the road, but here's what I know now.

The star of the show for midwinter production by far was **Tokyo bekana**. It germinates fast, grows even faster, and lasts through four cuttings without trying. It has a frilly light green leaf reminiscent of some leaf lettuces, but unlike them, it is a masterful presence in late December, when I can really use such a friend.

The winner for looks was **ruby streaks**. It's deeply lobed and frilly like mizuna but is more like the red Russian kale in substance. It also offers color indicative of its well-given name—pretty in the mix. This one, however, was also the slowest grower of the bunch, so it will not feature heavily in the midwinter mix. Ruby streaks comes on more quickly after the first cutting and looks just as good in its fourth harvest as it did for the first. I'll continue experimenting with this one to see how it does in fall and spring.

In 2008–2009, I decided to try red varieties of sorrel, komatsuna, and pac choi. I also tried an Italian dandelion and a couple of other new greens I saw in the catalogs. There are more varieties of Asian greens coming out each year, and they are well-suited to the winter greenhouse. I hope to have a few new ones to try each season. I remember wondering in my first year of production if I would get bored of growing mostly greens all winter long, but there are so many to try—and new varieties introduced in the seed catalogs each year—that boredom is the least of my worries. It's fun to see what comes up with each new test.

A Suggested Planting Schedule

I do not have a perfect script that can be copied to ensure maximum production, since such variables as cloud cover, weather (in and out of the greenhouse), and temperature conspire to ensure that nothing will ever be the same from year to year. That said, here's a rough outline of what I plant and when. You can use this as a guide and tweak your own schedule based on the results you get in your own greenhouse.

Late August and Early September

Start soil block plantings of broccoli and broccoli raab. Do the same with pac choi, Chinese cabbage, chard, and kale, along with more broccoli and raab as insurance, in case the first planting fails. If you end up not needing those backups, you can always eat the seedlings—they make a tasty little salad, and it doesn't feel wasteful to kill them if you eat them.

Mid-September

In planters, plant lettuce mix, cress, minutina, claytonia, mizuna, and arugula. You will really have to watch the arugula and mizuna. Place those planters on the bottom tier of the harnesses to keep them away from the warmer temperatures up high. It can get plenty hot in the greenhouse in September, so make sure you're venting on warm days and keeping fans going so the plants don't fry. Heat stress will shorten the life of an arugula or mizuna planting. The other varieties listed can handle the heat better than these two.

When I begin seeding planters in mid-September, I get 8 planters of greens going each weekend until all of the harnesses are filled. Eight is the maximum number of planters I can fit onto the 2 germination mats I use. Eight planters usually takes 2 batches of soil mix with some left over. I vary what's planted each week, allocating at least 4 of the planters for fast-growing varieties and at least 2 for slower-growing varieties, like claytonia or minutina.

October

Early in the month, the transplants for the raised beds are ready to go in. It's okay that they're still babies even though our CSA deliveries start in mid-October, because there are so many fresh veggies being harvested from outdoor garden beds. You can plant a lot of pac choi and Chinese cabbage transplants in one 3-foot-wide bed. For the small pac choi I grow, I plant 4 across the bed and space those mini rows about 6 inches apart. The Chinese cabbage is 3 plants per row. Broccoli is more of a pig about space, so I plant it in 2 rows, with the transplants alternating as they go down the row like this:

X X X X X X
  X X X X X X

That spacing also allows me to intercrop in the broccoli rows. In the spaces between broccoli plants I can tuck in pac choi and know that those plants will be harvested before the broccoli gets so big it shades them out. This helps me get over the annoyance I feel about broccoli taking up so much space in the greenhouse. The combined planting looks like this:

X x X x X x X x X x X
  X x X x X x X x X x X

In the October planters, I go for the arugula and mizuna but also include mustards like Osaka purple and red giant. I usually include a planter of tatsoi, komatsuna, or vitamin green with each week's planting, knowing they won't mature as quickly as the arugula or mizuna. I seed some other slow growers too, like chard, kale, or bull's blood beets.

For example, in the second week of October, I might seed 3 planters of arugula, 2 of mizuna, 2 of vitamin green, and 1 of bull's blood beets. In the third week of October, I might plant 3 mizuna, 2 tatsoi, 2 red giant, and 1 each of kale and cress. You get the idea. As soon as the sprouted seedlings put out their first true leaves, they go into the harnesses, and another round of planters is filled.

November

Now I'm into full swing with the mustards, and it's time to add that fabulous mid-season grower, Tokyo bekana. All the others previously mentioned are in the mix, too. This is my least sunny month of the winter season, so I pay close attention to the growing abilities of the varieties that are maturing at this time. A stretch of three or more weeks of mostly cloudy days is not uncommon at this time, and it does slow down the growing process. It also ensures that nobody is going to suffer from heat stress, so arugula and mizuna come into their own when harvested now.

December

By now, the harnesses are filled with planters, and the greenhouse is at full production. Variety selection is the same as with November at this point. Keep an eye out for the September planters that are showing signs of being done. Most will give you at least three harvests, but after that they will grow more slowly or sparsely or just look spent—especially the arugula. Take those planters out to the compost and dump them. These emptied planters can be washed and refilled with soil to keep the process going.

Also in December, as raised beds are harvested clean of their vegetables, new flats of transplants can be started to replace what has been taken. This means giving the flats some room on the germination mats so those seedlings can germinate quickly. But the timing works out well because planters are only needed when they are swapped out with the spent ones. Six planters a week is sufficient and gives room for 2 flats of transplants.

January

In January, I plant more Tokyo bekana because it reaches harvest size so quickly. Mizuna does well now, too, as do the other mustards, tatsoi, komatsuna, and vitamin green. The greenhouse is closed up tight and has been for a while because of the cold. Keep an eye out for signs of moisture trouble: moss growing on the soil surface or fungal problems with plants. Strategically placed fans help keep the air circulating, and I sometimes lightly hoe the soil surface in the raised beds to keep moss at bay if it gets a hold somewhere. If a planter has meltdown—my term for fungus that seems to melt a portion of the greens in a planter—out it goes. I dump it in the outdoor compost rather than risk having a problem spread.

Of course, on a sunny January day, it is wonderful to be able to open up the door to the packing shed and let some of that warm humid air hydrate the shed and garage. Oh, and let's not forget the real joy of January—sitting in the greenhouse working on the ol' journal while basking in humid warmth and sunlight!

February

Come February, I can really feel the difference in the longer days and the effect the increasing light has on the plants. Early in the month, I will plant hakurei turnips in any empty places in the raised beds, or maybe some icicle radishes. This is also the time to reintroduce lettuces into the planter rotation. A given planting might include 3 Tokyo bekana, 2 lettuce mix, 2 mizuna, and 1 vitamin green. I stop planting arugula because it bolts so quickly in the hotter weather of the March greenhouse.

February is also the month when I start planting the first of the seedlings that will go outdoors. Both celeriac and leeks benefit from the early start that a greenhouse can give them, so I get them going around Valentine's Day. Once they are up and going, they get placed on the shelf units above the germination mats.

March

March marks the end of seeding planters. Our CSA deliveries end in mid-April, so the only planters now are of lettuces that will get cut once for shareholders and then go outside to supply us with some early salad greens. I place spent planters of mustards and cole crops such as tatsoi on a picnic table near the greenhouse. When they go to flower, their small blossoms provide early food for emerging insects. I've read that beneficial insects like these kinds of flowers, so I invite them to my gardens early and give the last of the greenhouse planters a final purpose. They sometimes even continue growing and flowering where they've been dumped in the compost bin.

Back in the greenhouse, more flats of garden transplants take up space where sprouting planters and raised beds of other crops once reigned. Vents are open again, and the first breath of chilly spring helps moderate the rising temps in the greenhouse.

April

Finally, in April, the main work of the greenhouse is complete. The last of the planters and broccoli plants find their way to the compost pile. The greenhouse seems naked without the harnesses full of productive planters. The raised beds are now home to flats of thriving transplants and containers full of annual flowers that will grace the yard in summer. The place has an entirely different character in April compared to midwinter. By this time, I am eager again to expand into the larger spaces outdoors. But it is nice to know that even when winter's final nasty fits sweep in on a spring day, it is still warm and inviting in the greenhouse, where there's always something that needs doing. The final duty of this great growing space is to give young tomato and pepper plants a haven while they wait for the outdoor soil to be sufficiently warm for their roots. After that, the winter greenhouse sits empty and ignored until it's time to solarize and get ready for another autumn of activity.

Infinite Diversity in Infinite Combination

We know that we've found only the keys to this system, not all the doors they can unlock. We look forward to hearing about all the greenhouses that other peoples' ideas and situations will bring forth. Please share your questions, ideas, and adventures in winter growing with us at our online discussion site, http://gardengoddessnetwork.ning.com.

Resources

A Short Reading List

Listed here are the key books we read, highlighted, memorized, frequently referenced, and depended on. You are invited to absorb their wisdom and apply it according to your own needs.

Building

All of these books are out of print, but they are all also worth the hunt.

James C. McCullagh, editor. *The Solar Greenhouse Book.* **Emmaus, Pa.: Rodale Press, 1978.**

> This book includes useful tables on things like how many BTUs can be picked up at different times of day and estimates of heat loss, as well as sun path diagrams. It demonstrates different styles of passive solar greenhouses and explains different kinds of construction materials.

Edward Mazria. *The Passive Solar Energy Book: A Complete Guide to Passive Solar Home, Greenhouse and Building Design*. Emmaus, Pa.: Rodale Press, 1979.

> This text also illustrates different design options for a passive solar greenhouse. If you can get your hands on a copy, check out page 211. To our surprise, we found that Mazria shows a greenhouse with active rock heat storage similar to our own design. This book came out in 1979, so it radiantly demonstrates that all our good ideas are not new ones.

Spruille Braden III. *Graphic Standards of Solar Energy*. Boston: CBI Publishing, 1977.

> This one is full of graphs and formulas and maps of the United States that show prevailing winds in summer and winter, and wind patterns around various building shapes and . . . you get the idea. If you're into this sort of thing, this book's a gold mine.

Growing Produce

These are the main books I relied on for our first year of production. Some of what they told me worked, and some of it didn't. I adjust my methods with each year's experience, and you'll do the same. Don't be afraid to experiment. Always set aside a part of your growing space to try something new. You will create your own new story by recycling the experience of others. Remember to record what you learn each year so you can incorporate that invaluable information into future plans. Your journal is the best book you'll have.

Sandie Shores. *Growing and Selling Fresh-Cut Herbs*. Second edition. Batavia, IL.: Ball Publishing, 2003.

> If you are at all interested in growing herbs to sell, this is your book. Shores does a terrific job of taking the reader step by step through the process, from research to management, of starting a fresh herb business. Her advice on setting up a business and marketing the product is transferable to anyone who wants to

sell vegetables. She also gives useful information on managing a greenhouse in winter. This book is not easy to find because it is out of print, but do some web searching and check local used bookstores, and maybe you'll score a copy. It's worth looking for.

Anna Edey. *Solviva: How to Grow $500,000 on One Acre and Peace on Earth*. Vineyard Haven, Mass.: Trailblazer Press, 1998.

The ideas for passive solar design and winter production in this book are unique. The most important idea we garnered from Edey was to use planters to increase production space. She used other materials, in a different setup, but the results are similar. If we lived on a farm instead of in town, we would try out her idea about using animals to help provide heat for the greenhouse.

Other Greenhouses

Shane Smith. *Greenhouse Gardener's Companion: Growing Food and Flowers in Your Greenhouse or Sunspace*. Revised and expanded edition. Golden, Colo.: Fulcrum Publishing, 2000.

This is an excellent book for anyone new to greenhouse growing. It's Carol's favorite book on this topic. Smith covers the fundamentals of passive solar greenhouse design and the basics of growing both vegetables and ornamentals in a winter greenhouse. He offers useful planting schedules for a host of veggie varieties and goes for the organic solution first for any problems. He's also got a great website full of information and resources: www.greenhousegarden.com.

Eliot Coleman. *Four-Season Harvest: Organic Vegetables from Your Home Garden All Year Round*. White River Junction, Vt.: Chelsea Green, 1999.

Coleman is familiar to most organic farmers who have ventured into season extension using hoop houses. Carol used his crop variety suggestions as a knowledge base for Garden Goddess Produce. While not all of Coleman's information on winter hoop-house production is applicable to a passive solar greenhouse, he does base his methods on a principle we completely agree with: keep it simple. The less stuff you have to plug something in or burn fuel to produce, the more sustainable your system is. In his case, he went with unheated hoop houses with extra covering to protect crops on nights that dip below freezing. That just isn't going to work here on the Minnesota prairie in January, but the principle is still a sound one.

Take it down to the simplest system possible. That's why we're not pumping more heat into our greenhouse or installing an expensive lighting system to crank out mediocre tomatoes. Why do that when other crops will grow beautifully and happily with no added heat or light at all?

Coleman is also good at laying out the basics of organic farming practices. He has a website you can visit: www.fourseasonfarm.com.

Eliot Coleman. *The Winter Harvest Handbook: Year-Round Vegetables Using Deep Organic Techniques and Unheated Greenhouses.* White River Junction, Vt.: Chelsea Green, 2009.

This manual focuses entirely on Coleman's winter production system, which uses high tunnels with rows of crops that get extra protection from the cold through the use of floating row covers. His added design feature on these hoop-house structures is a skid base that allows the farmer to move the structure from one spot to another. If you want more details on that kind of growing method—which is definitely a season extension technique where we live—then check out this manual. Coleman provides all the details, including preferred seed varieties, tools, and equipment. This manual can be purchased at his website, www.fourseasonfarm.com

Seeds, Bugs, Equipment

Favorite Seed Catalogs

Before you buy from any seed company, do a little research at www.garden watchdog.com (rerouts to http://davesgarden.com/products/gwd/). This is a website where folks can go to submit or read reviews (thumbs up and thumbs down) of many seed catalog companies. If you look at the collected complaints and praise for a particular company, you soon get an idea of what you can expect. Some companies have customers who are consistently happy with seed quality but have terrible luck with transplants. Other companies give great customer service over the phone but never answer e-mails. Some are notorious for taking way too long to get a shipment in the mail.

If you can settle on a company that can satisfy most of your needs, you'll save on shipping costs. I usually get about 90 percent of what I want from one trustworthy source. I get the special varieties they don't carry from as few other sources as possible, to save on postage and handling. Sometimes I check around to see if friends are doing any ordering so we can join forces and split the shipping costs. This pays off when you're buying as much seed as I do.

Below are the companies I depend on. There are many to choose from, and you'll develop your own preferences. I know I can rely on these businesses to help me succeed with my own. No seed, no CSA.

Johnny's Select Seeds

955 Benton Avenue
Winslow, ME 04901
877-564-6697
www.johnnyseeds.com

This is the company I rely on most heavily. Their commercial growers' catalog has almost everything I need, in the quantities I want. You get little useful information on seed packs themselves; all that stuff is in the catalog, so don't throw it away after placing your order because you'll want to refer back to it later. Johnny's also has a great assortment of good quality garden tools, and ordering through their website is fairly easy.

Pinetree Garden Seeds

PO Box 300

New Gloucester, ME 04260

207-926-3400

www.superseeds.com

I've been buying from this little company for a lot of years now. I was first introduced to them by a friend who loved their small packs of seed (which also cost less). Fewer seeds per pack makes you feel like you can afford to try more varieties (at least that's the way it always works with me). The winter radish seed I buy from them does better than any others I've purchased from other suppliers. Since I know I'm going to be buying from Pinetree every year for that reason alone, I look for new stuff to try.

If you're also interested in things like making herbed vinegars or soaps or candles, or even growing plants for dyeing fiber, Pinetree's catalog has what you need.

Kitazawa Seed Co.

PO Box 13220

Oakland, CA 94661-3220

510-595-1188

www.kitazawaseed.com

Kitazawa's terrific catalog has a great selection of Asian greens and other Asian vegetables. It provides good details about the varieties and some interesting traditional Asian recipes to try. In addition to the crops I grow in the winter greenhouse, this catalog has inspired me to try some new varieties in my gardens outdoors as well, including gobo and many kinds of radishes.

Bugs

GreenMethods.com

93 Priest Road

Nottingham, NH 03290

603-942-8925

www.greenmethods.com

They're the only beneficial insect company I've used, and that's because I've had such great support from them—and great bugs. Beneficial insects are their world. They have an amazing, informative website for those of us new to the magic of beneficial insects and what they can do for us in the greenhouse. Shipping these insects is not cheap, so it's a good idea to find a supplier who is happy to help you select the right critters for your needs. These friendly folks also called me after each shipment arrived to make sure everything was okay. That's all I needed to keep their website and phone number close at hand.

Equipment

FarmTek

1440 Field of Dreams Way

Dyersville, IA 52040

1-800-476-9715

www.growerssupply.com

I love these folks—and not just because they're based in my original home state of Iowa. Getting their catalog is like reliving that feeling I had as a kid when the Sears Christmas catalog showed up. One of everything, Santa! Most of the greenhouse goods we find in the fancy home and garden and conventional greenhouse catalogs can be found through FarmTek for less money. They have items like good circulation fans at many sizes for any need. The list goes on and on, but it's all in there. I can spend a lot of pleasant time looking through that catalog. It gives me ideas.

We ordered our twin-wall polycarbonate sheets from them, along with their medium-sized greenhouse equipment package. The package deal saved us some money, and everything was delivered by truck on the day and time promised. I've also had great help from them over the phone when I've had questions. They treat me like an important customer even though I'm not a megafarmer. They're helpful, knowledgeable, and friendly—all good attributes in a company I plan to use again and again in the years to come. If I ever win the lottery, they are going to be very happy people.

Your Local Farm Store, Hardware Store, or Plant Nursery

We are firm believers in supporting local businesses. We want to have these stores nearby when we need them. In a rural area, that means giving them as much business as possible, even if things cost a little more than they do at some big-box home store farther away. What we've also found as a result of frequenting the local businesses is much better service. They actually help you.

Acknowledgments

We owe our current adventure in passive solar greenhouse production to the earlier efforts of researchers, farmers, and writers who've shared their experiences and wisdom from the 1970s to the present. We learned from the best information and ideas we could find and created our own combination of those concepts.

Our thanks to Shane Smith, Eliot Coleman, and Anna Edey for their useful information about simple but effective ways to grow food in the winter months.

In our region, we are blessed with a blossoming local foods community. Their support, encouragement, guidance, and belief in us and our venture were critical to its success.

Thanks to Audrey Arner and Richard Handeen at Moonstone Farm for reigniting the pioneer spirit at Richard's family homestead over thirty years ago. You helped foster a vibrant network of inspired land stewards.

Thanks also to Kay and Annette Fernholz at EarthRise Farm. You are true mentors. May we continue to learn from each other for many years to come.

To Terry VanDerPol, Amy Bacigalupo, and all the staff from the Land Stewardship Project, deepest thanks for the vital work you do to keep the foodland healthy and sustainable. Without LSP's Farm Beginnings course, we simply wouldn't have made this dream a reality.

We also owe Ruth Ann Karty sincere thanks for being Carol's Small Business Development Center counselor. Ruth Ann, you were the first, the best, the ever true cheerleader with business savvy and common sense that still keeps me steady.

A special thanks to Ben Winchester. He's the one who kept saying, "You guys gotta write a book about this stuff!" And he tirelessly urged us to make it happen. Ben, you were right. Thanks for being that persistent voice of encouragement.

Our editor, Ann Delgehausen, earns a special thanks for guiding us through the journey of developing a book that would serve the goal of helping others learn from our model. You took a couple of babes through the woods there, Ann, and taught us a lot along the way.

Also, a great big thanks to the board of directors of the University of Minnesota, Morris, West Central Sustainability Partnership for awarding us the grant that got this book off the ground. The Partnership is the Johnny Appleseed of our region, and many trees are taking root thanks to its vision and commitment to rural communities.

The last and largest dose of gratitude goes to our winter share members. You signed on for the ride when we weren't sure where the turns and twists would be. We couldn't have done it without you and your commitment to support local farmers and food.

THE Northlands Winter Greenhouse MANUAL

A Unique, Low-Tech Solution to Vegetable Production in Cold Climates

Carol Ford &
Chuck Waibel

You are welcome to contact Chuck and Carol

to have them speak at your event or consult for your project.

newworld@fedteldirect.net *or* **320.734.4669**

www.gardengoddessenterprises.com

Chuck Waibel and Carol Ford

The Garden Goddess Greenhouse has been under the watchful eyes and devoted care of Carol Ford and Chuck Waibel since 2005. Chuck designed the structure, and Carol figured out what to grow and when. They combine their fresh produce in winter with storage crops they raised the previous summer and fall for their CSA share members. Carol and Chuck attend conferences and workshops throughout the northlands to teach others what they've learned about winter vegetable production. Both are also musicians and writers and are active in local community organizations in the prairie village of Milan, Minnesota.

ENVIRONMENTAL BENEFITS STATEMENT

Garden Goddess Produce saved the following resources by printing the pages of this book on chlorine free paper made with 100% post-consumer waste.

TREES	WATER	ENERGY	SOLID WASTE	GREENHOUSE GASES
10 FULLY GROWN	**4,680** GALLONS	**0** MILLION BTUs	**0** POUNDS	**0** POUNDS

 Calculations based on research by Environmental Defense and the Paper Task Force.
Manufactured at Friesens Corporation